Valcárcel Cases · Ríos Castro

Who's Who in Analytical Chemistry

Springer-Verlag
Berlin Heidelberg
GmbH

Miguel Valcárcel Cases · Angel Rios Castro

Who´s Who in Analytical Chemistry

Europe

Springer

Prof. Miguel Valcárcel Cases
Dr. Angel Ríos Castro

Departamento de Química Analítica y Ecología
Facultad de Ciencias
Universidad de Córdoba
Avda. San Alberto Magno, s/n
14004 Córdoba
Spain

e-mail: qa1meobj@lucano.uco.es
qa1ricaa@uco.es

ISBN 978-3-642-62617-3

CIP data applied for

Die Deutsche Bibliothek - CIP-Einheitsaufnahme
Who`s who in analytical chemistry : Europe / M. Valcárcel Cases, A. Ríos Castro -
Berlin ; Heidelberg ; New York ; Barcelona ; Hong Kong ; London ; Milan ; Paris ; Singapore ;
Tokyo : Springer, 2002

ISBN 978-3-642-62617-3 ISBN 978-3-642-56354-6 (eBook)
DOI 10.1007/978-3-642-56354-6

http://www.springer.de

Originally published by Springer-Verlag Berlin Heidelberg New York in 2002
Softcover reprint of the hardcover 1st edition 2002

Typesetting: Dataconversion by author
Cover-design: Künkel + Lopka, Heidelberg
Printed on acid-free paper SPIN: 10725953 52 / 3020 hu - 5 4 3 2 1 0 -

Preface

The former Working Party, and currently Division, of Analytical Chemistry of the Federation of European Chemical Societies made the decision in 1996 to produce this book, which lists a large number of European analytical chemists together with their addresses and fields of interest. The principal aims of this book are to promote mutual knowledge and co-operation among them and to allow young chemists and private enterprises to choose a specific research group according to their analytical research lines and specialities.

Through our national delegates, we distributed a comprehensive questionnaire with no restrictions all over Europe. The response varied markedly among countries. We summarized the information received from each analytical chemist, which was confirmed on an individual basis. This was a time-consuming process, so we are aware that not every single European analytical chemist is listed and that some facts may by now be outdated. However, we believe that this first edition will be of invaluable help in order to promote Analytical Chemistry in Europe and remain one of the most salient outputs of the Division of Analytical Chemistry.

Finally, we should like to acknowledge the assistance received from Springer–Verlag (particularly from Mr Peter Enders) and the invaluable technical support of Mr J.M. Membrives.

M. VALCÁRCEL
A. RÍOS

CÓRDOBA, MAY 2002

Countries

Contents

Austria

Belgium

Brazil

Bulgaria

China

Czech Republic

Denmark

Estonia

Finland

France

Germany

Greece

Hungary

Ireland

Israel

Italy

Norway

Poland

Romania

Russia

Slovak Republic

Spain

Sweden

Switzerland

Turkey

Personalities

Austria

BENETKA, Edmund, Mag.
Austrian Research Centre
Seibersdorf "Chemical Analysis"
2444 Seibersdorf
Phone: +43 2254 780355
Fax: +43 2254 74060
E-Mail: benetka@zdfzs.arcs.ac.at

Domains of application:
Analysis of water, soils, plants, sewage sludges, food, air, human samples, metals, plastic

Applied methods:
ICP, AAS, GC, GC-MS, HPLC, IC, potentiometry

Membership:
EURACHEM

BUCHBERGER, Wolfgang, Dr.
Analytical Chemistry
University of Linz
Altenbergerstrasse 69
4040 Linz
Phone: +43 732 2468724
Fax: +43 732 2468679
E-Mail:
wolfgang.buchberger@jk.uni-linz.ac.at

Domains of application:
Organic trace analysis; pesticide residue analysis; environmental analytical chemistry; analytical chemistry for industrial product control

Applied methods:
GC; HPLC; capillary electrophoresis; mass spectrometry; hyphenated techniques; electroanalytical techniques; molecular spectroscopy

Membership:
GÖCh; ASAC; GDCh

Awards:
Feigl Prize of the Gesellschaft Österreichischer Chemiker

BUCHMEISER, Michael, Mag. Dr. rer.nat.
Institut für Analytische Chemie und Radiochemie
University of Innsbruck
Innrain 52
6020 Innsbruck
Phone: +43 512 5075184
Fax: +43 512 5072677
E-Mail:
michael.r.buchmeiser@uibk.ac.at

Domains of application:
Synthesis of new stationary phases for: solid phase extraction (SPE) of metal ions, HPLC (chiral chromatography, ion chromatography), monolithic stationary phases

Applied methods:
Metathesis polymerisation (ROMP, ADMET); Atom-transfer radical polymerisation (ATRP) HPLC; GC; SPE; GPC; AAS; ICP; FT-IR; UV-VIS; NMR; DSC; MS, BET, X-ray

Membership:
GÖCh; GDCh; ACS

Awards:
Prof. Ernst Brandl Preis (1998)

DICKERT, Franz L., Prof., Dr.
Institute of Analytical Chemistry
University of Vienna
Währingerstr. 38
1090 Vienna
Phone: +43 14277-52317
Fax: +43 1 4277-9523
E-Mail: franz.dickert@univie.ac.at

Private address:
Am Ölberg 161/Haus 2
3400 Klosterneuburg

Domains of application:
Chemical sensors for the detection of halogenated and aromatic hydrocarbons; voc analysis; oil quality; process analysis; air analysis; drinking water quality

Applied methods:
Chemical sensors; host-guest chemistry; molecular imprinting; application of IDC; ChemFET; SAW and QMB devices; spectroscopy, chemometry, pattern recognition

Membership:
GDCh; GÖCh

DOIBER, Rudolf, Dipl.-Ing., Dr. techn.
HBLVA für Chemische Industrie
Rosensteingasse, 79
1170 Vienna
Phone: +43 732 4861489
Fax: +43 732 4890359
E-Mail: rudolf.doiber@eunet.at

Private address:
Gablenzgasse 91/28
1150 Vienna

Domains of application:
pollution analysis (air, soil water); food analysis (preservants, colours, additives); analysis of technical products (automotives, plastics); analysis of fermentation products and byproducts enzymatic analysis

Applied methods:
Spectroscopic methods (UV/VIS, IR, AAS, ICP); chromatographic methods (GC, HPLC, IC); electrochemical (potentiometric, coulometric, voltammetric)

DUCHACZEK, Hubert, Dipl.-Ing., Dr.
Department Environmental Analytical Chemistry and Process Engineering
Voest-Alpine Stahl Linz GmbH
Voest-Alpine Str. 3
4031 Linz
Phone: +43 70 658275733
Fax: +43 70 69803496

Domains of application:
Analysis of air, gas, water, soil, waste, dust; analytical and technological assessment of lubricants; analysis of cool tar products

Applied methods:
VDI; ÖNORM; DIN; EPA; inhouse methods

Awards:
Certificate Accreditation based on accreditation law BGBI. Nr. 468/1992, decree Z1.92 710/27-IX/20/94; GAZ-Certificate: GAZ-Accreditation of Test Laboratory based on the Standards, EN 45001, ISO 9001

FRIEDBACHER, Gernot, Dipl.-Ing., Dr.
Institute of Analytical Chemistry
Vienna University of Technology
Getreidemarkt 9/151
1060 Vienna
Phone: +43 1 5880115110
Fax: +43 1 58801 15199
E-Mail:
gernot.friedbacher@tuwien.ac.at

Domains of application:
Development of in-situ characterization methods of materials with different scanning probe microscopy techniques; investigation of surfaces and surface processes with high spatial resolution

Applied methods:
Scanning probe microscopy; scanning tunneling microscopy; scanning force microscopy; atomic force microscopy and related techniques; electron probe x-ray microanalysis

Membership:
Commision V.2 on Microchemical Techniques and Trace Analysis; IUPAC; Austrian Society for Analytical Chemistry; GDCh

Awards:
Proanalytica awarded by the Faculty of Sciences of the Vienna University of Technology; Fritz Feigl Preis Awarded by the Austrian Society for Analytical Chemistry

---

FRÖHLICH, Johannes, Assist. Prof., Assoc. Prof., Dipl.-Ing., Dr.
Institute of Organic Chemistry
Vienna University of Technology
Getreidemarkt 9
1060 Vienna
Phone: +43 1 588015009
Fax: +43 1 5866931
E-Mail: jfroehli@pop.tuwien.ac.at

Private address:
Arbeitsgasse 50
1050 Vienna

Domains of application:
Organic synthesis; novel reagents; structure elucidation and synthesis of natural products; development of novel pharmaceuticals; heterocyclic chemistry; key-intermediates for material sciences

Applied methods:
1D/2D-NMR-spectroscopy; specific NMR-soltware simulation and separation methods; NMR-data bases and data archiving; HPLC-methods

Membership:
GDCh; GÖCh; ISHC; Österr. Pharmazeutische Gesellschaft

FUCHS, Dietmar, M.Sc., Dr.
University of Innsbruck
Fritz-Pregl-Strasse 3
6020 Innsbruck
Phone: +43 512 5073519
Fax: +43 512 5072865
E-Mail: dietmar.fuchs@uibk.ac.at

Private address:
Tiergartenstrasse 1/9
6020 Innsbruck
Domains of application:
Chemistry and biochemistry of pteridines; neopterin as tool in clinical laboratory diagnosis; tryptophan metabolism, parameters of immune activation; cytokines

Applied methods:
HPLC; ELISA; RIA; molecular biological techniques

Membership:
Clinical Immunology Society; Austrian Society of Allergy and Immunology, Austrian AIDS-Society, Austrian Society for Clinical Chemistry; Austrian Society for Biochemistry; International Society of Pteridinology

Awards:
Austrian AIDS Award; Award of the Austrian Society for Clinical Chemistry

---

GANN, Michael Oliver, Dipl.-Ing.
Institute for Building Biology and Ecology
Landstrasser Hauptstrasse 67
1030 Vienna
Phone: +43 1 7133793
Fax: +43 1 7120997
E-Mail: ibo@ping.at

Private address:
Pokornyg 31/6
1190 Vienna

Domains of application:
Research for healthy dwellings and living; investigation of building materials regardins their ecological and building-biological features; analytical investigation of indoor pollutants and xenobiotics; analytical method development

Applied methods:
GC; HPLC; UV-VIS

GRASSERBAUER, Manfred, Dipl.-Ing., Dr. techn.
Institute for Analytical Chemistry
Vienna University of Technology
Getreidemarkt 9/151
1060 Vienna
Phone: +43 1 5880115103
Fax: +43 1 5880115199
E-Mail: sek151@email.tuwien.ac.at

Domains of application:
Inorganic, speciation and organic trace analysis; micro and surface analysis; materials analysis (metals, semiconductors, ceramics, glass); environmental analysis (water, air, soil, plants); quality assurance (certification of reference materials)

Applied methods:
AAS; IC; GC; HPLC; MS; IR; MIP; CZE; EPMA; SIMS; AFM

Membership:
Austrian Society for Analytical Chemistry; Austrian Chemical-Physical Society; IUPAC; FECS

Awards:
Fritz Feigl Award of Austrian Society for Analytical Chemistry; Fritz Pregl Award of Austrian Academy of Sciences; Theophilus Redwood Lecture Award of Royal Society of Chemistry; Richard Kieffer Award of International Journal of Hard and Refractory Mat.

HASLINGER, Ernst, Dr. Phil.
Institute for Pharmaceutical Chemistry
University of Graz
Universitätsplatz 1
8010 Graz
Phone: +43 316 3805366
Fax: +43 316 3809846
E-Mail:
ernst.haslinger@kfunigraz.ac.at

Private address:
Jochbergengasse 26
1210 Vienna

Domains of application:
NMR-spectroscopy; structure-elucidation of natural products, mainly glycosides; synthesis of terpenoids

Applied methods:
NMR; GC-MS

Membership:
GÖCh; GDCh; Neue Schweizerische, Chemische Gesellschaft; International Society of Magnetic Resonance

Awards:
Forschungspreis of Sandoz Forschungsinstitut, 1979

HELLIGER, Wilfried, Dr. Phil.
Institute of Medical Chemistry and Biochemistry
Fritz-Pregl-Strasse 3
6010 Innsbruck
Phone: +43 512 5073521
Fax: +43 512 5072876
E-Mail: wilfried.helliger@uibk.ac.at

Private address:
Innrain 125

6020 Innsbruck

Domains of application:
Analysis of histones, histone subtypes and histone modifications; investigations concerning the biological function of histone proteins

Applied methods:
Electrophoresis; HPLC; capillary electrophoresis; protein sequencing

---

HOFER, Ferdinand, Dr., Dipl.-Ing.
Research Institute for Electron Microscopy
Graz University of Technology
Steyrergasse 17
8010 Graz
Phone: +43 316 8738820
Fax: +43 316 811596
E-Mail: f705hofe@mbox.tu-graz.ac.at

Domains of application:
Electron microscopy: SEM, TEM, STEM, EFTE; micro and nanoanalysis of materials, catalysts, biological tissue, thin films

Applied methods:
Electron microscopy: SEM, TEM, STEM, EFTEM; micro- and nanoanalysis; x-ray spectrometry (EDXS, WDXS); electron energy-loss spectrometry (EELS)

Membership:
GÖCh; Österr. Gesellschaft f. Elektronen- Mikroskopie; American Microbeam Analysis Society; European Microbeam Analysis Society, Materials Research Society

Awards:
Kardinal-Innitzer Förderungs Preis, 1989; Josef-Krainer-Würdigungspreis, 1994

HUTTER, Herbert, Dr., Dipl.-Ing.
Institute for Analytical Chemistry
Vienna University of Technology
Getreidemarkt 9/151
1060 Vienna
Phone: +43 1 588014849
Fax: +43 1 5867813
E-Mail: h.hutter@tuwien.ac.at

Domains of application:
Development of a system for recording and processing of ion micrographs for the three-dimensional characterization of a solid state material and application for the study of the distribution of trace elements in high performance materials

Applied methods:
SIMS

Awards:
Pro Analytica Awarded by the Faculty of Sciences of the Vienna University of Technology

---

JIROVETZ, Leopold, Mag. Pharm., Dr. rer. nat.
Pharmaceutical Chemistry
University of Vienna
Althanstrasse 14
1090 Vienna
Phone: +43 1 3133618641
Fax: +43 1 31336771
E-Mail:
jirovetz@speedy.pch.univie.ac.at

Domains of application:
Analysis of synthetic pharmaceutics drugs; analysis of natural products (essential oils); analysis of drugs from biological matrix

Applied methods:
GC; GC-MS; DI-MS;
NMR

Awards:
Herba Award, 1992; Czedik-

Austria

Eysenberg Award, 1994

KANDLER, Wolfgang, Dipl.-Ing. Dr.
IFA-Tulln
Konrad-Lorenz-Str. 20
3430 Tulln
Phone: +43 227266280406
Fax: +43 2272 66280 403
E-Mail: kandler@ifa-tulln.ac.at

Private address:
Alter Ziegelweg, 37-39/36
3430 Tulln

Domains of application:
Environmental analysis; quality assurance in environmental analysis; proficiency testing in environmental analysis

Applied methods:
IC; spectrophotometry; FAAS; ETAAS; HG-AAS; CV-AAS; SPE; HPLC; GC-MS; GC-ECD

KENNDLER, Ernst, Prof.
Institute for Analytical Chemistry
Währingerstr. 38
1090 Vienna
Phone: +43 1 4277 52305
Fax: +43 1 4277 9523
E-Mail: ernst.kenndler@univie.ac.at

Domains of application:
Bioanalytical chemistry; environmental chemistry; pharmaceutical chemistry

Applied methods:
Capillary electrophoresis; capillary chromatography

Awards:
Fritz Feigl Award (Austrian Society of Anal. Chem.); Fritz Pregl Award (Austrian Academy of Sciences)

KLAMPFL, Christian, Dr., Mag.
Analytical Chemistry
University of Linz
Altenbergerstrasse 69
4040 Linz
Phone: +43 732 2468722
Fax: +43 732 2468679
E-Mail: christian.klampfl@jk.uni-linz.ac.at

Domains of application:
Organic trace analysis; environmental analytical chemistry; analytical chemistry for industrial product control

Applied methods:
High performance liquid chromatography; capillary electrophoresis mass spectrometry; hyphenated techniques; molecular spectroscopy

Membership:
GDCh; Österr. Gesellschaft für Analytische Chemie

KNAPP, Günter, Dipl.-Ing., Dr. techn.
Institute of Analytical Chemistry, Micro- and Radiochemistry
Graz University of Technology
Technikerstrasse, 4
8010 Graz
Phone: +43 316 8738300
Fax: +43 316 8738304
E-Mail: knapp@analytchem.tu-graz.ac.at

Private address:
Sorgerweg 16, 8047 Graz

Domains of application:
Development of instrumentation and methods for sample preparation and element preconcentration; development of miniaturized RF-plasmas; development of methods for trace element determination by means of ICP-MS

Applied methods:
AAS; ICP-OES;ICP-MS; IC; GC; GC-MS; HPLC; microwave assisted sample preparation

Membership:
Austrian Chemical Society(GÖCh); ASAC; German Chemical Society (GDCh); A.M.S.E; Rotatory Club

Awards:
Honorary Doctor of the University of Veszprem, Hungary

KNOGLINGER, Johann Franz, Mag. rer.nat., Dipl.-Ing., Dr. techn.
Lenzing AG
Werkstrasse 1
4860 Lenzing
Phone: +43 7672 7012833
Fax: +43 7672 94046
E-Mail: lagknj@lenzing-ac.co.at

Private address:
Wiessen 8
4910 Ried

Domains of application:
Development of measuring and testing methods for fiber and textile industry

Applied methods:
Spectroscopy (UV, VIS, NIR, IR); gravimetry (for mixture content determination); electrochemical methods (conductimetry; voltametry; amperometry; potentiometry); microwave and untrasound for special measuring application

Membership:
GÖCh

KOPP, Brigitte, Prof. Dr.
Institut für Pharmakognosie
University of Vienna
Althanstr. 14
1090 Vienna
Phone: +43 1 313368367
Fax: +43 1 31336772
E-Mail: brigitte.kopp@univie.ac.at

Private address:
Standgasse 10
1180 Vienna

Domains of application:
Research; education; isolation; structure elucidation and analysis of natural compunds (i.e. plant material)

Applied methods:
HPLC; GC; CC; DCCC; CE; GC-MS

Membership:
Society for Medical Plant Research; Österr. Gesellschaft für Phytotherapie; Phytochemical Society of Europe

Awards:
Theodor Domes Presis; Herba Preis

KRSKA, Rudolf, Dipl.-Ing., Dr.
Center for Analytical Chemistry
IFA-Tulln
Konrad Lorenz Str. 20
3430 Tulln
Phone: +43 2272 6628040
Fax: +43 2272 662804
E-Mail: krska@ifa1.boku.ac.at

Domains of application:
Analysis of mycotoxins; pesticides; PAHs; OCPs; chlorinated hydrocarbons; elemental analysis matrix cereals, ester, soils; organisation of interlaboratory comparison tests; development of analytical methods; analysis of carbohydrates (structure), fats

Applied methods:
HPLC-UV; DAD; FLD; HPLC-PBI-MS; HPIC-PAD; SEC-LALLS; CZE; GC-ECD/FID/NPD/MS; ET-

AAS; FAAS/CV-AAS/HG-AAS; FTIR; SPE; TDC; SFE; MAE; elementary analyzer for C, H, N, S. TLC

Membership:
GÖCh; Austrian Society for Analytical Chemistry

Awards:
Pro Analytica 1993

KUHNERT-BRANDSTÄTTER, Maria, Dr. rer.nat., Mag. Pharm.
Institute of Pharmacognosy
University of Innsbruck
Innrain 52
6020 Innsbruck
Phone: +43 512 5075302
Fax: +43 512 5072939
E-Mail: Pharmakognosie@uibk.ac.at

Private address:
Greid 12
6082 Patsch

Domains of application:
Teaching pharmacognosy; research work on medical plants and pharmaceuticals; special line: polymorphism of drugs and its relevance to analysis, solubility and stability of preparations

Applied methods:
Thermomicroscopy; differential scanning calorimetry; thermogravimetry; IR-spectroscopy; FTIR microspectroscopy; FT Raman spectroscopy; x-ray diffractometry; different chromatographic methods

Membership:
American Microchemical Society; Hungarian Pharmaceutical Society; Austrian Pharmaceutical Society; Österreichisches Ehrenkreuz für Wissenschaft und I. Klasse

Awards:
Fritz Pregl Award of the Austrian Academy of Sciences; Carl Mannich Medal of the German Pharmaceutical Society

---

LÄMMERHOFER, Michael, Dr.
Institute of Analytical Chemistry
University of Vienna
Währingerstr. 38
1090 Vienna
Phone: +43 1 427752323
Fax: +43 1 42779523
E-Mail:
michael.laemmerhofer@univie.ac.at

Domains of application:
Chiral separations with LC, CE, CEC, crystallization stereoselective analysis and bioanalysis of chiral drugs; preparative separation and isolation of chiral compunds; development of stereoselective separation systems

Applied methods:
Liquid chromatographic methods; capillary electrophoretic and capillary electrochromatographic methods; mass-spectrometry hyphenated with LC/CE/CEC (LC-MS, CE-MS, CEC-MS)

Membership:
GÖCh; Österreichische Pharmazeutische Gesellschaft; The QSAR and Molecular Modelling Society

---

LAMCHE, Herbert R., Dipl.-Ing., Dr. techn.
Bender Co.
Dr. Boehringer, 5-11
1121 Vienna
Phone: +43 1 80105996
Fax: +43 1 80105366
E-Mail: lamche@bender.co.at

Private address:
Friedhofsallee 457
2534 Alland

Domains of application:
Pharmaceutical R&D
Applied methods:
Immunoassay (RIA, ELISA);
bioassay (celluler assays)

Membership:
Clinical Ligand Assay Society;
European Ligand Assay Society

LATKOCZY, Christopher, Dr.
Institute of Chemistry
University of Agricultural Sciences -
BOKU Vienna
Muthgasse 18
1190 Vienna
Phone: +43 1 360066050
Fax: +43 1 360066059
E-Mail: clatkocz@mail.boku.ac.at

Domains of application:
High resolution-ICPMS;
environmental analysis; soil
analysis; isotopic ratios
determination in environmental
samples, otolith analysis, laser
ablation ICP-MS

Applied methods:
HR-ICPMS; HPLC; Laser
LUV266X, microwave digestion;
clean room (class 10000)

Membership:
GÖCh; ASAC; ACS; RSC

LEITNER, Erich, Dipl.-Ing.,
Dr.techn.
Department of Food Chemistry and
Technology
Graz University of Technology
Petersgasse 12/2
8010 Graz
Phone: +43 316 8734522
Fax: +43 316 8736971
E-Mail: erich.leitner@tugraz.at

Domains of application:
Aroma and off-flavour analysis in
foodstuff and packaging materials

Applied methods:
GC-MS, GC olfactometry, sensory
analysis

Membership:
GÖCh; ASAC

LENDL, Bernhard, Dipl.-Ing., Dr.
techn.
Institute of Analytical Chemistry
Vienna University of Technology
Getreidemarkt 9/151
1060 Vienna
Phone: +43 1 5880115140
Fax: +43 1 58801-15199
E-Mail: blendl@fbch.tuwien.ac.at

Domains of application:
Process analysis; chemical analysis;
vibrational spectroscopy;
miniaturized analysis systems;
sensor development

Applied methods:
FIA; sequential injection analysis;
FTIR-spectroscopy; Raman
spectroscopy

Membership:
GÖCh, ASAC

Awards:
Pro Analitica, 1996

MÜLLER, Ingun-Astrid, Dipl.-
Biochem.
Hydrologische Untersuchungsstelle
Salzburg
Lindhafstr. 5
Salzburg
Phone: +43 662 43325732
Fax: +43 662 48790442

Private address:
Thrimegger Str. 8A
5020 Salzburg

Domains of application:
Water analysis (drinking and waste water, ground and surface water); ecotoxicology; biochemistry; consulting; hygiene, hydrobiology, geohydrology, geology, planning
Applied methods:
GC; HPLC; spectroscopy (UV, IR, FL); IC

Membership:
GDCh; GÖCh

MURKOVIC, Michael, Assoc.Prof., Dr.
Department of Food Chemistry and Technology
Graz University of Technology
Petersgasse 12/2
8010 Graz
Phone: +43 316 8736495
Fax: +43 316 8736971
E-Mail:
michael.murkovic@tugraz.at

Domains of application:
Analytical food chemistry

Applied methods:
HPLC (UV, FLD, DAD, ECD)

Membership:
GÖCh; ASAC; Austrian Nutritional Society; AOCS

NIKIFOROV, Alexes, Prof. Dr.
Institute of Organic Chemistry
University of Vienna
Währingerstr. 38
1090 Vienna
Phone: +43 1 313672210
Fax: +43 1 313672280
E-Mail: niki@felix.orc.univie.ac.at

Domains of application:
Organic analysis of complex mixtures

Applied methods:
GC-MS; multivariate analysis

Awards:
Fritz-Pregl-Price 1984

PFANNHAUSER, Werner, Prof. Dr.
Department of Food Chemistry and Technology
Graz University of Technology
Petersgasse 12/2
8010 Graz
Phone: +43 316 8736470
Fax: +43 316 8736971
E-Mail:
werner.pfannhauser@tugraz.at

Private address:
Kreuzg. 79
1180 Vienna

Domains of application:
Food analysis (contaminants, ingredients, model reactions, antioxidation, substances, food colors, PAH, NO_2-PAH, glucosides, flavour analysis)

Applied methods:
GC/MS; SFE; GC/NPD,ECD; HPLC/DAD,FD; CE; electrophoretic + immunological methods; GPC; AAS

Membership:
FECS-DFC; Austrian Society of Nutrition

Awards:
Gold Medal of the Italian Society of Flavor Analysis

PFARL, Christian, Dipl.-Ing.
Agrolinz Melamin GmbH

Welser Str. 42
4060 Leonding
Phone: +43 732 69142587
Fax: +43 732 672106

Domains of application:
Environmental chemistry and residue chemistry (organic trace analysis)

Applied methods:
Chromatography (HPLC, GC); spectroscopy (MS, UV)

Membership:
GÖCh, ASAC

PROHASKA, Thomas, Dipl.-Ing., Dr. techn.
Institute of Chemistry
University of Agricultural Sciences
Muthgasse 18
1190 Vienna
Phone: +43 1 360066092
Fax: +43 1 360066059
E-Mail: prohaska@mail.boku.ac.at

Private address:
Einsiedlergasse
1050 Vienna

Domains of application:
Elemental trace analysis, isotope ratio measurements

Applied methods:
ICPMS; AAS; hydride generation, HLPC, GC;

PUXBAUM, Hans, Dipl.-Ing., Dr.
Institute for Analytical Chemistry
Vienna University of Technology
Getreidemarkt 9
1060 Vienna
Phone: +43 1 5880115170
Fax: +43 1 58801 15199
E-Mail:
hpuxbaum@mail.zserv.tuwien.ac.at

Private address:
Stuckgasse 10
1070 Vienna

Domains of application:
Environmental analytical chemistry with special emphasis on atmospheric gas, aerosol, cloud and precipitation chemistry

Applied methods:
IC; HPLC; GC; GC-MS; TOC; DC-FID; thermal analysis; combustion analysis; denuder techniques, trace gas analysis with chemiluminescence techniques

Membership:
ASAC; Society of Aerosol Research; American Geophysical Union; European Geophysical Society; Austrian Academy of Sciences

Awards:
Fritz Pregl Award (Austrian Academy of Sciences); Fritz Feigl Award (Austrian Society of Analytical Chemistry)

ROBIEN, Wolfgang,
Institute of Organic Chemistry
Währingerstr. 38
1090 Vienna
Phone: +43 1 427752119
Fax: +43 1 313652129
E-Mail:
wolfgang.robien@univie.ac.at

Domains of application:
NMR-spectroscopy; computer assisted structure elucidation; software developments

Applied methods:
NMR (1D, nD)

Membership:
GÖCh; GDCh

Austria

Awards:
Theodor Körner Preis

SCHMID, Erich R., Prof. Dr.
Institute of Analytical Chemistry
University of Vienna
Währingerstr. 38
1090 Vienna
Phone: +43 1 313672402

Private address:
Peter Jordan-Strasse 39
1190 Vienna

Domains of application:
Instrumental analysis; drugs spectrometry; UV/VIS-fluorescence; screening methods; applied fields

Applied methods:
Spectrometry; separation methods; screening tools

Membership:
GÖCh; GDCh; Österr. Gesellschaft für Analytische Chemie

Awards:
Feigl Prize

SONTAG, Gerhard, Dr.
Institute of Analytical Chemistry
University of Vienna
Währingerstr. 38, 1090 Vienna
Phone: +43 1 313672403
Fax: +43 1 3196312
E-Mail: gerhard.sontag@univie.ac.at

Domains of application:
Electroanalytical chemistry; biosensors; food chemistry

Applied methods:
Electroanalytical methods (voltammetry, polarography, amperometry); HPLC and FIA with electrochemical detectors; electrochemical biosensors

---

STINGEDER, Gerhard, Dipl.-Ing., Dr.
Institute of Chemistry
University of Agricultural Sciences (BOKU)
Muthgasse 18
1190 Vienna
Phone: +43 1 360066091
Fax: +43 1 360066059
E-Mail: gstin@mail.boku.ac.at

Domains of application:
Elemental ultratrace analysis under clean room condition

Applied methods:
ICP-MS; AAS, HPLC; IC; ICP-AES

STUPPNER, Hermann, Dr., Mag. Pharm.
Institut für Pharmakognosie
University of Innsbruck
Innrain 52
6020 Innsbruck
Phone: +43 512 3075307
Fax: +43 512 5072939
E-Mail:
hermann.struppner@uibk.ac.at

Domains of application:
Analysis of plant constituents; phytochemistry

Applied methods:
GC; HPLC; CE; HPLC/MS; CE/MS; TLC

VYCUDILIK, Walter, Dr.
Institute of Forensic Medicine
University of Vienna
Sensengasse 2
1090 Vienna
Phone: +43 1 4024051
Fax: +43 1 4052726
E-Mail:
gerichtsmedizin@univie.ac.at

Private address:
Knödelhüttenstrasse 25/1/3

1140 Vienna

Domains of application:
Detection of low molecular compounds in biological materials (legal and illicit drugs, metabolites, heavy metals)

Applied methods:
Mass spectrometry, chromatography and spectroscopic methods (IR, UV, AAS, XFA)

Membership:
Gesellschaft für Toxikolog. und Forens. Chemie; Deutsche Gesellschaft für Rechtsmedizin; International Association of Forensic Toxicologists

---

WINKER, Norbert, Dipl.-Ing., Dr.
Allgemeine Unfallversicherungsanstalt
Adalbert Stifter Strasse 65
1200 Vienna
Phone: +43 1 33111213
Fax: +43 1 33111218

Domains of application:
Determination of explosion parameters; determination of minimal ignition energy of dusts; diverse environmental analyses; investigations fo working places (hazardous dusts and chemical substances)

Membership:
Expert of European Commission Directorate V; Directorate XII; Member of the Scientific Committee for the post graduate course on Occupational Health ETH Zürich, International Social Security Association; International Commission on Occupational Health

Belgium

FRAUERWIESER, Gerhard, Dipl.-Ing., Dr.
European Commission
DG IA.C.2
rue D'Arcon 88
1040 Brussels
Phone: +32 2 2990216
Fax: +32 2 29640305
E-Mail:
gerhard.frauerwieser@dg1a.cec.be

Private address:
rue Des Minimes, 22
1000 Brussels

Domains of application:
Foreign economic reactions EU-TACIS- and phare-countries; trade relations EU-Russia; technical barriers to trade EU-Russia; standardization; certification; accreditation of testings; laboratories; mutual recognition agreements (MRAs); WTO-TBT-CODE

Membership:
VDI

LAMBERTS, Luc, Dr.
The University Namur (B)
Facultés Universitaires N.D. de la Paux
rue de Bruxelles, 61
5000 Namur
Phone: +32 81 724511
Fax: +32 81 724530
E-Mail: luc.lamberts@fundp.ac.be

Domains of application:
Organic electrochemistry; potentiometry stripping analysis; chiral separations; environment

Applied methods:
General electrochemical methods: ultramicroelectrochemistry and

Belgium

cyclic voltammetry; spectrochemical analysis; chromatographs separations

Membership:
Socié Royale de Chimie; Societé Chimique

MAIER, Eddie Arthur, Dr.
European Commission
Directorate General for Research
rue de la Loi, 200
1049 Brussels
Phone: +32 2 2956138
Fax: +32 2 2958072
E-Mail: eddie.maier@cec.eu.int

Private address:
rue de la Rive, 96
1200 Brussels

Domains of application:
R&D: instrumentation-QA-QC-certified reference materials; support to standardisation, antidoping and antifraud

Applied methods:
Organic-inorganic-microbiological methods

Membership:
IUPAC; AOAC, SFC, IAWQ

RODRÍGUEZ, Adela, Prof. Dr.
Institute for Research Materials and Measurements
Joint Research Centre
Retieseweg
2440 Geel
Phone: +32 14 571211
Fax: +32 14 581273
E-Mail:
adela.rodriguez@irmm.jrc.be
Private address:
Eegden, 11
2440 Geel

Domains of application:
Development of analytical methods in the fields of cosmetic products, food, and Environment. Research on life Sciences.

Applied methods:
Spectrometry gamma and alpha, atomic absorption, ICP-MS, HPLC-DAD-ED-MS/MS, capillary electrophoresis-DAD-MS; electrochemical methods high precision balances

Membership:
European Commision; Chairperson of the Working Group on Analytical Methods for Cosmetic Products

Brazil

ZAGATTO, Elias A.G., Prof. Dr.
Centro de Energia Nuclear na Agricultura
Universidade de Sao Paulo
Av Centenario 303
13416-000 Piracicaba SP
Phone: +55 19 4294650
Fax: +55 19 4294610

Private address:
R. Maçons 478
13405-040 Piracicaba SP

Domains of application:
Environmental analysis; food analysis; instrumentation; chemometrics

Applied methods:
Automated analysis of agronomical and environmental samples; flow analysis; spectrophotometry

Membership:
IUPAC; Commission U1-General Aspects of Analytical Chemistry

ARNAUDOV, Michael, Dr. D. Sci. Chem.
Laboratory of Molecular Spectral Analysis
University of Sofia
J. Bourchier Blvd. 1
1164 Sofia
Phone: +359 2 6256208
Fax: +359 2 9625438
E-Mail: marnaudov@chem.uni-sofia.bg

Private address:
Rakovsky sr. 122b
1000 Sofia

Domains of application:
Research institutions; analytical laboratories

Applied methods:
IR-spectroscopy; UV-vis spectroscopy

BARDAROV, Ventzislav,
Military Medical Academy
Toxicochemical Laboratory
Georgi Sofiisky Str. 3
1606 Sofia
Phone: +359 2 9172960
Fax: +359 2 958 1543
E-Mail: chroma@cblink.net

Private address:
Stephan Dimitrov Str. 16
1225 Sofia

Domains of application:
Medical and environmental applications of chromatographic analysis

Applied methods:
GC; HPLC; GC/MS

BONTCHEV, Panayot, Dr.
Department of Analytical Chemistry
University of Sofia "St. Kl. Ochridski"
J. Bourchier Blvd. 1
1164 Sofia
Phone: +359 2 628323
Fax: +359 2 9625438
E-Mail: prbontchev@chem.uni-sofia.bg

Private address:
Drujba 2, Bl. 263/B, app. 49
1582 Sofia

Domains of application:
Research institutions; R&D centres; industrial, analytical and control laboratories

Applied methods:
Emission spectroscopy; IR & UV-vis; AAS; EFE; magnetic measurements; TGA, EPR (ESR), ESCA (electron spectroscopy for chemical analysis), NMR.

Membership:
Alliance "Universities for Democracy"

Awards:
Highest Bulgarian Award in Chemistry "A. Zlatarov", 1985; Medal of Honour of Slovak Technical University, 1985; Cyril and Methodius Order, 1st Degree; Kliment Okhridski Medal, Highest Degree with a Blue Ribbon

---

CHRISTOVA TOPALOVA, Ivanka, Dr.
Faculty of Chemistry
University of Sofia "St. Kl. Ochridski"
J. Bourchier Blvd. 1
1164 Sofia
Phone: +359 2 6256367
Fax: +359 2 9625438

E-Mail: itopalova@chem.uni-sofia.bg

Private address:
Busludsha 39
1463 Sofia

Domains of application:
Analytical chemistry; organic chemistry; agrochemistry; food chemistry; environmental analysis

Applied methods:
HPLC; GC; GC-MS

DIMOV, Neno P., Prof., Dr.
Laboratory of Chromatography
Chemical Pharmaceutical Research Institute
Kliment Okhridsky, 3
1756 Sofia
Phone: +359 2 74321
Fax: +359 2 742163

Private address:
Kv. "Mladost" I, Bl. 66
1784 Sofia

Domains of application:
Analyses of drug substances and formulations, fermentation and natural products, waste water, quality assurance and control, QSAR and QSRR

Applied methods:
All kinds of chromatographic and optical spectral methods; electrochemical analytical methods; chemometrical methods

Membership:
ACS; FECS-DAC

FUTEKOV, Lybomir, Prof., Dr. Sci.
Department of Analytical Chemistry
University of Plovdiv
Avcenty Str. 49
4000 Plovdiv
Phone: +359 2 260061
Fax: +359 2 235049
E-Mail: futekov@ulcc.uni-plovdin.bg

Domains of application:
Inorganic analyses; organic analyses; environmental analyses; food analyses

Applied methods:
AAS; ICP; ICP-MS; chromatography; chemometrics

IVANOV, Kalintcho, Dr.
Analytical Chemistry
Higher Institute of Food and Flavour Industry
Maritza Blvd. 26
4002 Plovdiv
Phone: +359 2 44181226
Fax: +359 2 440102

Private address:
Pencho Slaveicov Str. 7
4002 Plovdiv

Domains of application:
Analysis of food and biologically active substances

Applied methods:
Chromatography; TLC; GC; HPLC; spectrophotometry

KOVACHEVA, Kerka, Dr.
Petrochemical Research Institute
Neftochim-EAD
8104 Bourgas
Phone: +359 56 45865
Fax: +359 56 45865
E-Mail: inn-neftochim@mcr1.poptel.org.uk

Private address:
Slaveikov 4
8005 Bourgas

Domains of application:
Oil-processing and petrochemical syntesis; synthesis of polymers, elastomers and fibres; corrosion protection; environmental protection; application of analytical methods in oil-processing, petrochemical and environmental analysis

Applied methods:
Fourier-transform IR spectroscopy; UV-vis spectroscopy; atomic spectrometry; sample preparation methods

---

KOVATCHEVA, Apostolova, Dr.
Analytical Chemistry
Higher Institute of Food and Flavour Industry
Maritza Blvd. 26
4002 Plovdiv
Phone: +359 32 603 683
Fax: +359 32 44 01 02

Private address:
Jagoda Str. 1
4004 Plovdiv

Domains of application:
Analytical Chemistry and instrumental methods of food analysis (chromatography, spectroscopy and electrochemical methods); analysis of food componentes; various additives and their influence on food quality

Applied methods:
Thin layer chromatography; gas chromatography; HPLC; spectral and electrochemical methods

---

LEKOVA, Katya, Dr., Assoc. Prof.
Department of Chemistry
Southwestern University of Blagoevgrad
Ivan Mihailov, 66
2700 Blagoevgrad
Phone: +369 7320767
Fax: +359 9540203
E-Mail: ai@mail.techno-link.com
Private address:
Bukston Complex, Bl. 10
1518 Sofia

Domains of application:
Environmental sciences; spectral and separation methods in organic and inorganic chemistry

Applied methods:
Chromatographic methods

MILINA, Roumiana, Dr.
Petrochemical Research Institute
Neftochim-EAD
8104 Bourgas
Phone: +359 56 45865
Fax: +359 56 45865
E-Mail: inn-neftochim@mcr1.poptel.org.uk

Private address:
Zorniza, 18
8000 Bourgas

Domains of application:
Oil processing and petrochemical synthesis; synthesis of polymers, elastomers and fibres; corrosion protection; environmental protection; application of analytical method in oil-processing, petrochemical and environmental analysis

Applied methods:
GC; HPLC; GC-MS; solid phase extraction; Fourier transform IR spectroscopy; UV-vis spectroscopy; atomic spectrometry

---

NEICHEVA, Anastasia, Dr.
Analytical Chemistry
Higher Institute of Food and Flavour Industry
Maritza Blvd. 26
4002 Plovdiv

Phone: +359 32 603 626
Fax: +359 32 440102

Private address:
48 Bratia Buckstone
4004 Plovdiv

Domains of application:
Analytical chemistry and instrumental methods of food analysis; analysis of food components-pesticide residues, vitamins and biologically-active substances

Applied methods:
TLC; GC; HPLC; spectrophotometry

NIKOLOVA-DAMYANOVA, Boryana, Dr.
Institute of Organic Chemistry with Center of Phytochemistry
Bulgarian Academy of Sciences
Acad. G. Bonchev, 9
1113 Sofia
Phone: +359 2 724817
Fax: +359 2 700225
E-Mail: bmd@orgchm.acad.bg

Domains of application:
Isolation; structural elucidation; chemical properties; biological activity of natural organic compounds of terrestrial and marine origin; methods for separation, purification and quantification of natural organic compounds

Applied methods:
Gas-liquid chromatography; high-performance liquid chromatography; thin layer chromatography

---

OBRECHKOVA, Danka, Assoc. Prof.,Dr.
Medical University of Sofia
Faculty of Pharmacy
2, Dunav Str.
1000 Sofia
Phone: +988 3142295
Fax: +987 9874

Private address:
Vaskresenie bl.9, ap.25
1373 Sofia

Domains of application:
Research and quality of natural and pharmaceutical products

Applied methods:
Spectrophotometry; TLC, HPLC, GL

Awards:
ACS; New York Academy of Sciences

PAPAZOVA, Donka, Dr.
Petrochemical Research Institute
Neftochim-EAD
8104 Bourgas
Phone: +359 56 45865
Fax: +359 56 45865
E-Mail: inn-neftpchim@mcr1.poptel.org.uk

Private address:
Lazur 2
8001 Bourgas

Domains of application:
Oil processing and petrochemical synthesis; synthesis of polymers, elastomers and fibres; corrosion protection; environmental protection; application of analytical methods in oil-processing, petrochemical and environmental analysis

Applied methods:
Gas chromatography; HPLC; GC-MS; solid phase extraction; Fourier-transform IR spectroscopy; UV-vis spectroscopy; atomic spectrometry

---

POPOV, Dimitre, Dr.
Analytical Chemistry

Higher Institute of Food and Flavour Industry
Maritza Blvd. 26
4002 Plovdiv
Phone: +359 32 603713
Fax: +359 32 440102
E-Mail: gpopv@hotmail.com

Private address:
Petar Chilev Str. 18, ap. 38
4003 Plovdiv

Domains of application:
Analysis of foods and biologically active substances, stabilisation and analysis of wine

Applied methods:
Chromatography; TLC; GC; HPLC; spectrophotometry

SIMEONOV PETROV, Stilijn, Mag. Pharm.
Sopharma
Pharmaceutical Company
Iliensko Chausse Str. 16
1220 Sofia
Phone: +359 2 385531
Domains of application:
Analytical chemistry

Applied methods:
HPLC; GC; TLC

Awards:
International Symposium of Perkin Elmer, Third Prize

SPASSOV, Stefan L., PhD, Dr.Sci.
Institute of Organic Chemistry with Center of Phytochemistry
Bulgarian Academy of Sciences
Acad. G. Bonchev, 9
1113 Sofia
Phone: +359 2 7334264
Fax: +359 2 700225
E-Mail: spassov@orgchm.bas.bg

Private address:
Lajos Kossuth St. 9
1606 Sofia

Domains of application:
Organic chemistry; bioorganic chemistry; chemistry of natural products; chemistry of fuels; organic spectroscopy; qualitative, quantitative and structural analysis of organic compounds; theoretical organic chemistry

Applied methods:
Chemical methods of synthesis and analysis; spectroscopic methods (UV-vis, IR, multinuclear NMR, MS, CD); various chromatographic methods (GC, HPLC); theoretical methods (quantum chemistry, molecular mechanism)

STOEV, Georgi, Assoc. Prof.
Central Laboratory of Veter. Control and Ecology
Iskarsko Shousse Blvd. 5
1528 Sofia
Phone: +359 2 793023

Private address:
Mladost-1, 44
1184 Sofia

Domains of application:
Control of drugs, pesticides and heavy metal in food

Applied methods:
GC-MS; HPLC

Awards:
Book Practical Capillary Gas Chromatography

TASHKOV, Wladislav, Dr.
National Centre of Hygiene
Nestorov Str. 15
1431 Sofia
Phone: +359 2 5812743

Fax: +359 2 9581277

Private address:
Kupenite Str. 1, 1618 Sofia

Domains of application:
Hygienic control; food control

Applied methods:
GC; headspace

Membership:
New York Academy of Sciences

Awards:
Awards from the Health Ministery for excellent elaboration

China

JIN, Qunhan, Prof.
Jilin University
Jiefang Road 119
130023 Changchun
Phone: +86 431 8922331
Fax: +86 431 8922161
E-Mail: qhjin@public.cc.jl.cn

Private address:
Suite 101, Liutiao Road 3
130061 Changchun

Domains of application:
Analytical chemistry; microwave plasma chemistry; microwave chemistry; environmental analytical chemistry

Applied methods:
Plasma spectrometry

Membership:
Society of Applied Spectroscopy; Chinese Chemical Society; Chinese Electronics Society; Chinese Society of Analytical Instruments.

Awards:
State Award of Innovation of China; 5 Awards of Progress in Science and Technology form the State Commission of Education of China; an Award from the Ministry of International Trade of China and 3 Awards from the Government of Jilin Province.

Czech Republic

BAREK, Jiri, PhD, C.Sc., FRSC., C.Chem., EurChem.
Department of Analytical Chemistry
Charles University
Albertov 2030
12843 Prague 2
Phone: +420 2 21952295
Fax: +420 2 24913538
E-Mail: barek@natur.cuni.cz

Private address:
U svobodarny 9
19000 Praha 9

Domains of application:
Electroanalytical chemistry; HPLC

Applied methods:
Differential pulse polarography/voltammetry; adsorptive stripping; voltammetry; HPLC with electrochemical detection

Membership:
FECS-DAC; Czech Chemical Society, Royal Society of Chemistry

BOCEK, Petr, Prof., RNDr. Dr. Sci.
Institute of Analytical Chemistry
Academy of Sciences of the Czech Republic
Veveri 97
61142 Brno
Phone: +420 5 7268182
Fax: +420 5 41212113
E-Mail: bocek@iach.cz

Domains of application:
Analytical chemistry; separation methods; electrophoresis; capillary isotachophoresis; CZE

Applied methods:
Capillary isotachophoresis; capillary zone electrophoresis; combination of isotachophoresis and zone electrophoresis

Awards:
Price of the Czechoslovak Academy of Sciences, 1991

CHMELIK, Josef, Dr.
Institute of Analytical Chemistry
Academy of Sciences of the Czech Republic
Veveri 97
61142 Brno
Phone: +420 5 7268214
E-Mail: chmelik@ipm.cz

Domains of application:
Separation and characterization of polymers and particles; micropreparation of proteins and living cells

Applied methods:
Capillary electrophoresis and isoelectric focussing; field-flow fractionation

Awards:
1989: Award of Czechoslovak Biochemical Society

CHURACEK, Jaroslav, Prof., Dr. Sci.
Department of Analytical Chemistry
University of Pardubice
Nam. Cs. Legii 565
53210 Pardubice
Phone: +420 40 603 7086
Fax: +420 40 603 7068
E-Mail: Jaroslav.Churacek@upce.cz

Private address:
Pichlova 2533
53002 Pardubice

Applied methods:
Identification and separation or organic compounds-using chromatographic techniques, trace analysis, environmental analysis and supercritical fluid extraction. 172 publications, 1020 references after SCI and 16 monographs.

Membership:
Central Committee of Czech Chemical Society; Chairman of Pardubice Branch of Chemical Society. Chromatographic Society, Assembly of Academie of Czech Republic, Editorial Boards of Chemical Journals Chemicke Listy

Awards:
Hanus Medal of Czechoslovak Chemical Society 1980, Medal of University of Chemical Technology, Pardubice, 1983 and Honour Member of Czech Chemical Society 1985.

DEDINA, Jiri, RNDr., C.Sc.
Laboratory of Trace Element Analysis
Institute of Analytical Chemistry, ASCR
Videnska 1083
14220 Phaha 4
Phone: +420 2 4752490
Fax: +420 2 4752499
E-Mail: dedina@biomed.cas.cz

Domains of application:
Investigation of generation of volatile compounds and of their atomization in flame atomizers, quartz tubes and in graphite furnaces for atomic absorption and atomic fluorescence spectrometry

---

FLIEGER, Miroslav, Dr.
Institute of Microbiology
Academy of Sciences of the Czech Republic
Videnska 1083
14220 Phaha 4
Phone: +420 2 4712241
Fax: +420 2 4752347
E-Mail: flieger@biomed.cas.cz

Private address:
Poupetova 12/1331
17000 Praha 7

Domains of application:
Analytical chemistry; mycotoxicology; in vitro production of secondary metabolites; immunomodulatory effect of natural substances; development of methods for mycotoxin analysis of microbial products; development of methods for pharmaceutical analysis

Applied methods:
Chromatography (HPLC; GC; analytical, preparative); electrophoresis (analytical, preparative, native); isoelectric focussing capillary zone electrophoresis; chiral HPLC; chiral CZE; submerged fermentation or ergot alkaloids; pharmaceutical analysis

Membership:
American Society of Pharmacology; Biotechnological Society; American Chemical Society; Society for Leukocyte Biology; Czech Chemical Society; National Geographic Society

---

JANDERA, Pavel, PhD; Dr. Sci.
Department of Analytical Chemistry
Faculty of Chemical Technology
Nam. Cs. Legii 565
53210 Pardubice
Phone: +420 40 603 7023
Fax: +420 40 603 7068
E-Mail: pavel.jandera@upce.cz

Domains of application:
Organic analysis-spectroscopic and chromatographic methods and trace analysis; chromatographic-theory; instrumentation and applications; theory of gradient elution in HPLC; retention mechanism; calibration; prediction and optimization

Applied methods:
HPLC; HPLC/MS; capillary electrophoresis; UV-Vis and IR spectroscopy

Membership:
IUPAC; Editorial Board of "Journal of Chromatography", "Egyptian Journal of Analytical Chemistry"; reviewer of "Chromatographia", "Analytical Chemistry"

Awards:
Tswett Foundation Award, 1988

KARLICEK, Rolf, Prof.
Department of Analytical Chemistry
Faculty of Pharmacy
Heyrovskeho 1203
50005 Hradec Kralove
Phone: +420 49 5210718
Fax: +420 49 5210718
E-Mail: karlicek@faf.cuni.cz

Private address:
Hradec Kralove 11
50326 Stefanikova

Domains of application:
Flow methods; automation of analytical methods; pharmaceutical analysis

Applied methods:
Flow injection analysis; fluorimetry; solid-phase extraction on-line in flow system UV-vis spectrophotometry

---

KASICKA, Vaclav, RNDr.; C. Sci.
Institute of Organic Chemistry and Biochemistry
Academy of Sciences of the Czech Republic
Flemingovo, 2
16610 Prague 6
Phone: +420 2 20183239
Fax: +420 2 24310090
E-Mail: kasicka@uoch.cas.cz

Private address:
Renoirova 619/6
15200 Praha 5

Domains of application:
Development of instrumentation and methodology of electromigration separation methods and their application to analysis and preparation of biomolecules. Structure-mobility and structure-activity relationships of biologically active peptides and their derivatives

Applied methods:
Analytical capillary electromigration techniques: zone electrophoresis, isotachophoresis, electrokinetic chromatography; preparative free-flow electromigration techniques: zone electrophoresis, isotachophoresis; bioaffinity sorption, electrodesorption

Membership:
International Council of Electrophoretic Societies, representative (chairman) of the Chromatography and Electrophoresis Group of the Czech Chemical Society.

Awards:
Award of Czech Chemical Society in Analytical Chemistry, 1988.

KLEPARNIK, Karel, Eng. C. Sci.
Institute of Analytical Chemistry
Veveri 97
61142 Brno
Phone: +420 5 7268122
Fax: +420 5 41212113
E-Mail: klep@ipm.cz

Domains of application:
Separation methods; electrophoresis; analysis of DNA

Applied methods:
DNA restriction fragments length polymorphism; single stranded DNA conformation polymorphism; PCR analysis; DNA sequencing; pulsed field capillary electrophoresis

---

KOMAREK, Karel, Dr., Eng. C. Sci.
Faculty of Chemical Technology
Department of Analytical Chemistry
Nam. Cs. Legii 565
53210 Pardubice
Phone: +420 40 512111
Fax: +420 40 514530
E-Mail: komarek@hib.upce.cz

Domains of application:
Analytical chemistry; organic analysis

Applied methods:
GC; CGC; CGC-MSD; derivatization; reaction GC

NEMCOVA, Irena, PhD (Rerum Naturalium Doctoris -Czech, not international title)
Assoc. Prof.
Department of Analytical Chemistry
Faculty of Sciences
Albertov 2030
12840 Prague 2
Phone: +420 2 21952303
Fax: +420 2 24913538
E-Mail: inemcova@natur.cuni.cz

Private address:
Malenicka 1789
14800 Praha 4

Domains of application:
Analysis of biologically important substances: drugs in pure and dosage forms, surfactants in environment, heavy metals in plant and animal tissues (to remove: speciation of metals)

Applied methods:
UV-vis molecular and atomic absorption spectroscopy; flow-through methods (flow injection analysis, continuous flow analysis)

Membership:
Management Committee of COST Action

PACAKOVA, Vera, Dr., Assoc. Prof.
Department of Analytical Chemistry
Charles University
Albertov 2030
12840 Prague 2
Phone: +420 2 21952300
Fax: +420 2 24913538
E-Mail: pacakove@.natur.cuni.cz

Private address:
Na strzi 57
14000 Praha 4

Domains of application:
Separation sciences: gas, liquid chromatography, capillary electrophoresis; applications to biologically important substances: peptides, proteins, allergens; environmental analysis: triazines, combustion products; development of new analytical methods

Applied methods:
GC; pyrolysis gas chromatography; liquid chromatography with different detectors; capillary electrophoresis

Membership:
Czech Chemical Society, European Peptide Society

Awards:
Award for the best textbook in 1990 "Electrochemical Measurements in Flowing Liquids"

POLASEK, Miroslav, Assoc. Prof.; Dr.
Faculty of Pharmacy
Charles University
Heyrovskeho 1203
50005 Hradec Kralove
Phone: +420 49 5067304
Fax: +420 49 521002
E-Mail: polasek@faf.cuni.es

Domains of application:
Automation of analytical methods; pharmaceutical analysis with use of flow injection and electro-migration technique; electrochemical biosensors

Applied methods:
UV-vis spectrophotometry; fluorimetry; amperometry; other electroanalytical techniques; isotachophoresis; capillary zone electrophoresis

---

SEVCIK, G.K., Assoc. Prof.; Dr.
Faculty of Sciences
Charles University
Albertov 2030
12840 Prague 2
Phone: +420 2 1952296
Fax: +420 2 4913538
E-Mail: sevcik@prfdec.natur.cuni.cz

Domains of application:
Separation techniques; electrochemistry; some of an analytical methods

Applied methods:
Gas chromatography; multidimensional techniques; chemometrics; QSRR-QSPR; computer aided modelling of analytical processes

Awards:
R&D 100 Award

SOLICH, Petr, Assoc. Prof.; Dr.
Department of Analytical Chemistry
Faculty of Pharmacy
Heyrovskeho 1203
50005 Hradec Kralove
Phone: +420 49 5067294
Fax: +420 49 5210718
E-Mail: solich@faf.cuni.cz

Private address:
Lhota Liboany 146
50327 Stefanikova

Domains of application:
Pharmaceutical analysis; flow methods; dissolution studies of drugs

Applied methods:
Flow injection analysis; spectrophotometry; fluorescence

STULIK, Karel, Prof., Eng., Dr. Sci.
Faculty of Sciences
Charles University
Albertov 2030
12840 Prague
Phone: +420 2 21951111
Fax: +420 2 24913538
E-Mail: stulik@prfdec.natur.cuni.cz

Domains of application:
General analytical chemistry; gas and liquid chromatography, electrophoresis; electroanalytical chemistry; molecular spectroscopy; AAS; application to the environment clinical and pharmaceutical fields

Membership:
IUPAC

VEJROSTA, Jifi, Dr.
Institute of Analytical Chemistry
Academy of Sciences of the Czech Republic
Veveri 97
61142 Brno
Phone: +420 5 7268175
Fax: +420 5 41212113
E-Mail: vejrosta@iach.cz

Domains of application:
SFE; SFC; pressurized fluid extraction (PFE), fluid-phase equilibria in water supercritical carbon dioxide- solute systems, instrumentation development

Applied methods:
SFE; SFC; PFE,GC, GC-MS; HPLC

VENTURA, Karel, Assoc. Prof., Dr.
Department of Analytical Chemistry
University of Pardubice
Nam. Cs. Legii 565
53210 Pardubice
Phone: +420 40 603 7088
Fax: +420 40 603 7068
E-Mail: karel.ventura@upce.cz

Domains of application:
Analytical chemistry; organic analysis

Applied methods:
SFE;ASE; SPE; GC;GC/MS,HPLC

VESPALEC, Radim, Assoc. Prof., Dipl. Eng., Dsc.
Institute of Analytical Chemistry
Czech Academy of Sciences
Veveri 97
61142 Brno
Phone: +420 5 7268121
Fax: +420 5 41212113
E-Mail: vespalec@iach.cz

Domains of application:
Chiral separations; CZE investigation of analytical and physicochemical aspects of fast reversible interactions in solution; utilization of the dual nature of liquid chromatography including

characterization of solid surfaces and solid-liquid interfaces.

Applied methods:
CZE; HPLC; hyphenation HPLC - streaming current.

VYTRAS, Karel, Prof. Eng. Dr. Sci.
Department of Analytical Chemistry
University of Pardubice
Nam. Cs. Legii 565
53210 Pardubice
Phone: +420 40 6037512
Fax: +420 40 6037068
E-Mail: karel.vytras@upce.cz

Private address:
Dasicka 1767
53003 Pardubice

Domains of application:
Electroanalytical chemistry; instrumental analysis; Chemical Equilibria

Membership:
IUPAC

WURST, Milan, Dr.
Institute of Microbiology
Department of Biogenesis and Biotechnology of Natural Products
Videnska 1083
14220 Phaha 4
Phone: +420 2 4752647
Fax: +420 2 4752347
E-Mail: behal@biomed.cas.cz

Private address:
Ovalova 349/7
16000 Praha 6

Domains of application:
Analysis of mixtures of microbiological metabolites; separation, isolation, identification and determination of natural products

Applied methods:
Gas chromatography; HPLC; high performance capillary electrophoresis

Membership:
Czechoslovak Society for Microbiology; Czech Chemical Society

Awards:
Prize of Czechoslovak Academy of Sciences, 1981; Awards of Czechoslovak Academy of Sciences, 1987; Award of Czechoslovak Academy of Sciences, 1989

Denmark

CARLSEN, Lars, M.Sci., Dr. Sci.
Department of Environmental Chemistry
National Environmental Research Institute
Frederivs Borguej 399
4000 Roskilde
Phone: +45 46 301350
Fax: +45 46 301114
E-Mail: lc@dmu.dk

Private address:
Hyldeholm 4 Veddelev
4000 Roskilde

Domains of application:
Environmental chemistry; analytical chemistry; monitoring; reference laboratory activities; control analysis

Applied methods:
GC; GC/MS; LC; LC/MS

Membership:
Royal Society of Chemistry; American Chemical Society; Danish Chemical Society

Awards:
Accreditation (DANAK):

Environmental Analysis

CHRISTENSEN, Jytte Molin, M.Sci., Dr. Med. Sci.
Department of Chemistry and Biochemistry
National Institute of Occupational Health

Membership:
IUPAC; EUROLAB; Danish Accreditation; Danish Society for Occupational Medicine; Danish Society for Clinical, Chemical, Pharmacology; EURACHEM

GRON, Christian, M.Sci., Dr.
Department of Plantbiology and Biogeochemistry
Riso National Laboratory

Membership:
International Substances Society; European Association of Organic Geochemistry; Society of Environmental Toxicology and Chemistry

---

GUNDERSEN, Vagn, Bc. Sci.
Risoe National Laboratory
Plant Biology and Biogeochemistry Department
PO Box 49
4000 Roskilde
Phone: +45 46 774677
Fax: +45 42 370264
E-Mail: vagn.gundersen@risoe.dk

Domains of application:
Trace element analysis; analysis of organic micropollutants in soil and plants; wet oxidation of biomass; remediation of contaminated soil; analysis of degradation soil; analysis of degradation products of biomass after wet oxidation; dating of sediments

Applied methods:
ICP-MS; HRICPMS; ASS; GC; GC-MS/MS; IC; TOC; HPLC; XRFS; spectrometry

---

HALLGREN, Lars, M.Sci., Dr.
Novo Nordisk A/S
Analytical Biotechnology, R&D
Novo Alle; Bygn. IEP 10
2880 Bagsvaerd
Phone: +45 44 422574
Fax: +45 44 421492
E-Mail: lhal@novo.dk

Private address:
Aabrinken 191
2830 Virum

Domains of application:
Assay and instrumental development for process analytical chemistry (on-line, at-line and off-line)

Applied methods:
SIA/FIA; HPLC/FPLC; HPCE; centrifugal analyzers; manual methods

---

HANSEN, Elo Harald, Dr. Sci.
Department of Chemistry
Technical University of Denmark
Building 207
2800 Lyngby
Phone: +45 45 252346
Fax: +45 45 883136
E-Mail: ehh@kemi.dtu.dk

Private address:
Granasen 93
2800 Lyngby

Domains of application:
Analytical chemistry and corrosion; physical and biochemical chemistry; quantum chemistry; material and molten salt chemistry; structural chemistry

Applied methods:
Optical methods (UV/VIS; AAS, chemiluminescence, fluorimetry); electrochemical (potentiometry, ISE, amperometry, voltammetry, stripping methods, STM); flow injection analysis; automated methods; chemometrics

Membership:
WPAC/DAC; IUPAC

Awards:
Ole Romers Award; Ulrich Brinch's Award; G.A. Hagemann Gold Medal

---

HANSEN, Steen Honore, Dr. Sci.
Department of Analytical and Pharmaceutical Chemistry
Universitetsparken 2
2100 Copenhagen 0
Phone: +45 35 356777
Fax: +45 35 375376
E-Mail: shh@charon.dfh.dk

Domains of application:
Separation mechanisms in separation sciences (HPLC, CE); hyphenation of separation methods with spectroscopic techniques (LC-MS/MS, CE-MS/MS, LC-NMR-MS); drug metabolim; metabolism of trace elements

Applied methods:
Separation methods: HPLC, FPLC, CE, GLC; detection techniques: UV, fluorescence, MS, electrochemical, NMR

Membership:
Scandinavian Section of the Chromatographic Society

HOLCH, Klaus, M.Sci., Dr.
Institute of Food Chemistry and Nutrition
National Food Agency of Denmark
Morichoj Bygade 19
2860 Soborg
Phone: +45 39 696600
Fax: +45 39 660100
E-Mail: lst@lst.min.dk

Domains of application:
Food analysis; materials coming into contact with food

Applied methods:
HPLC; GC; ICP; HPLC-MS; GC-MS; ICP-MS; AAS; FIA; ELISA; microbiological assays; Fourier-Transform-IR

JENSEN, Ib, Bc. Sci.
Department of Applied Chemistry
Technical University of Denmark
Building 375
2800 Lyngby
Phone: +45 45 25 5422
Fax: +45 45 881770
E-Mail: ifakiij@unidhp.uni-c.dk

Private address:
Fasanvaenget 320
2980 Kohhedal

Domains of application:
Inorganic and analytical chemistry; physical chemistry; spectroscopy

Applied methods:
UV/VIS; IR; NMR; AAS; potentiometry

JENSEN, Ole Jorgen, M.Sci.Chem.
Radiometer Medical A/S
Akandevej 21
2700 Bronshoj
Phone: +45 38 273243
Fax: +45 38 272727
E-Mail: radforsk@ibm.net

Private address:
Dyssegardsvej 51B
2860 Soborg

Domains of application:
Clinical chemistry; instrumental analysis

Applied methods:
Electrochemical analysis with potentiometric sensor (ISE, pH); amperometric sensors (O2- and biosensors)

Membership:
International Society of Electrochemistry

JORGENSEN, Soren S., M.Sci., Dr.
Department of Chemistry
Royal Veterinary and Agricultural University
Thorvaldsensvej 40
1871 Frederiksberg
Phone: +45 35 282427
Fax: +45 35 282398
E-Mail: sstj@kvl.dk

Domains of application:
Food analysis; soil and water analysis; teaching analytical chemistry

Applied methods:
Atomic and molecular spectrometry; GC; HPLC; CE; MS; FIA

Membership:
IUPAC

KRYGER, Lars, Dr., M.Sci.
Department of Chemistry
Aarhus Univeristy
Langelandsgade 140
8000 Aarhus C
Phone: +45 79 423867
Fax: +49 86 196199
E-Mail: kry@kemi.aau.dk

Domains of application:
Research, development and education

Applied methods:
Construction, development and characterization of modified electrodes intended for selective and sensitive determination of traces of metallic species; stripping methods; assessment of the environmental impact of land reclamation

Membership:
Advisory Board of Talanta

RASMUSSEN, Allan, M.Sci., Dr.
Danish Meat Research Institute
Maglegaardsvej 2
4000 Roskilde
Phone: +45 42 361200
Fax: +45 42 364836
E-Mail: ajr@dmri.dk

Private address:
Rogeskovvej 3
4250 Fugleberg

Domains of application:
R&D equipment for analysis of meat quality

Applied methods:
Spectroscopy VIS-NIR

Estonia

HODREJARV, Helvi, Dr.
Tallinn Technical University
Ehitajate tee 5
0026 Tallinn
Phone: +372 6 202803
Fax: +372 6 202020
E-Mail: helvi@edu.ttu.ee

Private address:
Jaaroku tee 50
0020 Tallinn

Domains of application:
Environmental analytical chemistry;

Estonia

heavy metal; chemometrical data research; history of analytical chemistry

Applied methods:
AAS

Membership:
FECS-DAC

KALJURAND, Mihkel, Dr.
Institute of Chemistry
Tallinn Technical University
Ehitajate tee 5
0026 Tallinn
Phone: +372 2 202803
Fax: +372 2 536371
E-Mail: mihkel@argus.chemnet.ee

Domains of application:
GC; HPLC; chemometrics; AAS; application of computers in analytical chemistry for reduction of detection limits and monitoring time varying processes; using chemometrics for data processing

Applied methods:
GC; HPLC; AAS; chemometrics

Membership:
Editorial Board of "Journal of High Resolution Chromatography"

Awards:
Estonian National Science Award for Computer Chromatography, 1991

KOEL, Mihkel,
Institute of Chemistry
Akadeemia tee 15
0026 Tallinn
Phone: +372 2 536438
Fax: +372 2 536371
E-Mail: vmihkel@argus.chemnet.ee

Private address:
Ehitajate tee 46-33
0029 Tallinn

Domains of application:
Analytical chemistry

Applied methods:
Chromatography; spectrometry; chemometrics; supercritical fluid extraction

Membership:
Editorial Board of Journal "Critical Reviews of Analytical Chemistry" (USA)

Awards:
1991 Estonian State Award on Science (together with E Krillik and M. Kaljiurand)

KOPPEL, Elmar, Dr. Sci.
Institute of Chemical Physics
Tartu University
Javobi st. 2
2400 Tartu
Phone: +372 7 465263
Fax: +372 7 465264
E-Mail: ilmar@chem.ut.ee

Domains of application:
Analytical chemistry/instrumental methods of analysis; NMR; FT-IR; GC-MS; UV-vis; GC; HPLC; AES; AAS; OES; FT-ICR; non aqueous titration; chemical physics; bioorganic chemistry; theoretical chemistry; physical chemistry; organic chemistry

Membership:
ACS; RSC

Awards:
EX, US, Estonian, etc. Grants

LEPANE, Viia, M.Sci., Dipl. Chem.
Department of Inorganic and Analytical Chemistry
Tallinn Technical University
Ehitajate tee 5
0026 Tallinn

Phone: +372 6 202800
Fax: +372 6 202020
E-Mail: viia@edu.ttu.ee

Private address:
Nurmiku tee 12
0020 Tallinn

Domains of application:
Environmental analytical chemistry; stripping voltammetric determinations of heavy metals in natural waters and sediments; isolation and characterization of aquatic humic substances

Applied methods:
Stripping voltammetry; adsorption/partition chromatography; size exclusion chromatography; UV-vis spectroscopy

Awards:
Grants from Johnson Foundation, 1990; The Nordic Council of Ministers, 1994; The Knut and Alice Wattenberg Foundation, 1995

ORAV, Anne, Dr. Phil.
Ministry of Education
Institute of Chemistry
Akadeemia tee 15
0026 Tallinn
Phone: +372 2 536430
Fax: +372 2 536371

Private address:
Oismae tee 125-33
0035 Tallinn

Domains of application:
ion chromatography, SFE, GC, complex study of the thermodynamics of multicomponent systems and analytical general services of other departments (spectromertic and elementary analysis)

Applied methods:
Capillary gas chromatography

PAAMA, Lilli, Dr.
Institute of Chemical Physics
University of Tartu
Jakobi Street 2
2400 Tartu
Phone: +372 7 465265
Fax: +372 7 465264
E-Mail: peeter@chem.ut.ee

Private address:
Kalda tee 20-21
2400 Tartu

Domains of application:
Major factors affectin the line intensities of the elements with various exitation potential in inductively coupled plasma atomic emission spectrometry (ICP-AES); excitation mechanism in argon plasma, ion-atom ratios in inductively coupled plasma

Applied methods:
Inductively coupled plasma atomic emission spectrometry (ICP-AES); electrothermal atomic absorption spectrometry (ETAAS); thermal analysis and differential thermal analysis (TA-DTA); fourier transformed infrared spectroscopy (FTIR)

Awards:
Scholarships: International Science Foundation (ISF); Nordic Council of Ministers; European Training Foundation, Tempus Phare; Open Estonian Foundation

---

PEHK, Tonis, Dr. Sci.
Institute of Chemical Physics and Biophysics
Akadeemia tee 23
0026 Tallinn
Phone: +372 6 398319

Fax: +372 6 398393
E-Mail: pehk@kbyi.ee

Private address:
Trommi 6-21
0026 Tallinn

Domains of application:
NMR spectroscopy; mass spectrometry; chromatography; reaction mechanisms; catalysis; natural products; isotope effects; polymers; biopolymers; HT-superconductors

Applied methods:
NMR spectroscopy; MS spectrometry; chromatography

Awards:
Estonian Award for the Development of NMR Spectroscopy, 1972; Estonian Award for the Prostaglandin Research, 1987; P. Kogerman Memorial Medal on Chemical Research, 1991

VIITAK, Anu, Dr.
Tallinn Technical University
Ehitajate tee 5
0026 Tallinn
Phone: +372 6 202803
Fax: +372 6 202020
E-Mail: aruv@edu.ttu.ee

Private address:
Marrimetsa tee 25B
0026 Tallinn

Domains of application:
AAS; HPLC; analytical chemistry of microelements of human biomedia

Applied methods:
AAS; HPLC

Finland

IVASKA, Ari, Dr.
Laboratory of Analytical Chemistry
Abo Academy University
Piispankato, 8
20500 Turku-Abo
Phone: +358 2 2154420
Fax: +358 2 2154479
E-Mail: aivasha@abo.fi

Private address:
Vanha Harmeentic, 63b
20540 Turku

Domains of application:
Electroanalytical chemistry; process analytical chemistry; automatic analytical methods; environmental analysis; conducting polymers; analytical applications

Applied methods:
Electrochemistry (potentiometry, voltammetry, impedance measurement, electrochemical quartz microbalance); FTIR; ICP-MS

Membership:
Instrumental Society of America; The Electrochemical Society; International Society of Electrochemistry

---

KANKARE, Jouko, Dr.
Department of Chemistry
University of Turku
Vatselankatu, 2
20014 Turku
Phone: +358 2 3336710
Fax: +358 2 3336709
E-Mail: jouko.kankare@utu.fi

Domains of application:
Sensor technology, new materials for sensors, molecular electronics, conductive polymers, electrochemistry and electrochemical analysis, luminescence

Applied methods:
Various optical, electroanalytical and spectroelectrochemical methods

Membership:
International Society of Electrochemistry; The Electrochemical Society; ACS

KOMPPA, Veikko, Dr. techn.
VTT Chemical Technology
PO Box 1401
02044 VTT
Phone: +358 9 456560
Fax: +358 9 4567026
E-Mail: veikko.komppa@vtt.fi

Domains of application:
Organic analysis; chem. Metrology; synthesis; process clevel

Applied methods:
GC, MS, IR, NMR, HPLC, MALDI-TOF, ES-MS, HR-MS, mainly environmental, biotechnical and industrial org. Samples

Membership:
EURACHEM, AOAC, EUROMET, CITAC

KULMALA, Sakari, Dr.
Helsinki University of Technology
University of Turku
Kemistintie, 1
02015 Hut
Phone: +358 9 4512601
Fax: +358 9 462373
E-Mail: skari.kulmala@hut.fi

Domains of application:
Clinical analysis; point of care diagnostics; environmental analysis; general inorganic analysis

Applied methods:
Luminescence and electrochemical methods

Membership:
International Society of Electrochemistry; The Electrochemical Society

---

LAJUNEN, Lauri, Dr.
University of Oulu
Pentti Kaiteran Katu, 1
90401 Oulu
Phone: +358 8 5534071
Fax: +358 8 5544557
E-Mail: lauri.lajunen@oulu.fi

Private address:
Soutajantle, 24
90560 Oulu

Domains of application:
Trace elemental analysis by atomic absorption and plasma atomic emission methods

Membership:
IUPAC-V.6 titular member

MINKKINEN, Pentti, M. Techn.
Department of Chemical Technology
Lappeenranta University of Technology
Skinnarilankatu, 34
53851 Lappeenranta
Phone: +358 5 6212102
Fax: +358 5 6212199
E-Mail: pentti.minkkinen@lut.fi

Domains of application:
Education of chemical engineering

Applied methods:
AAS; ion chromatography; SEM, NIR; process and environmental analytical chemistry and process chemometrics; quality control of sampling and analytical measurements; properties of weak acids

Membership:
Member of Editorial Advisory Board

of CHEMOLAB

NIINISTO, Lauri, M.Sci., Dr. techn.
Laboratory of Inorganic and
Analytical Chemistry
Helsinki University of Technology
PO Box 6100
02015 Espoo
Phone: +358 9 4512600
Fax: +358 9 462373
E-Mail: lauri.niinisto@hut.fi

Domains of application:
Inorganic materials: synthesis and analysis; surface analysis; thermal analysis; luminescence studies

Applied methods:
XRF; AAS; SSMS; SIMS; XPS; luminescence spectroscopy; FTIR; TG; DTA; DSC; EGA-MS

Membership:
FECS; European Rare Farth Research Society

Awards:
Stanislao Cannizzaro Gold Medal of the Italian Chemical Society; Prize of the Association of Finnish Chemical Societies; The Chemists' Club Golden Award; Tallinn Technical University, doctor honoris causa

---

OSTERHOLM, Heidi, Techn. Lic., M.Sci.
Corporate Technology
Fortum Oyj/Neste Oyj
PO Box 310
006101 Porvoo
Phone: +358 204 503374
Fax: +358 204 503697
E-Mail: heidi.osterholm@neste.com

Private address:
Ostrav. 18
06100 Porvoo

Domains of application:
Catalysts, oil, polymers, resins

Applied methods:
x-ray diffraction, x-ray fluorescence spectrometry, temperature programmed desorption, N2-adsorption (AAS, ICP, GC, GC-MS, LC, NMR microscopy)

Membership:
ACS; SKS; SPPY

SIREN, Heli M.M., Dr.
VTT Chemical Technology
Biologintuja, 7
02044 VTT
Phone: +358 9 4565272
Fax: +358 9 4567026
E-Mail: heli.siren@vtt.fi

Private address:
Putousrinne
01600 Vantaa

Domains of application:
Environmental analysis; drug analysis; bioanalysts; instrument development; method development; on-line analysis (on-line process control)

Applied methods:
HPLC; GC; capillary electrophoresis (CE, CE-MS); sample preparation methods; UV/VIS spectrophotometry, and FL; mass spectrometry; solvent chemistry

Membership:
Member of Nordic Separation Science Society, Secretary in Finland

---

WASBERG, Mikael, Dr.
Laboratory of Inorganic Chemistry
Abo Akademi
Biskopsgatan, 8
20500 Abo

Phone: +358 2 2154421
Fax: +358 2 2154962
E-Mail: mwasberg@abo.fi

Domains of application:
Glass technology; combustion chemistry; materials chemistry; electrochemistry

Applied methods:
Voltammetry; quartz crystal balance (electrochemical); flow methods in electrochemistry

Membership:
International Society of Electrochemistry; Electrochemical Society; Association of Finnish Chemical Societies

France

ARTUSO, Thierry, Dr.
National Agency for Bererinary Medicinal Products
PO Box 203
35302 Fougeres
Phone: +33 2 99947860
Fax: +33 2 99947864
E-Mail: vafo30@calvacon.fr

Domains of application:
Control of the quality of the veterinary medicinal products; growting of marketing authorization for veterinary medicinal products

Applied methods:
HPLC (principally); titrimetric methods; Karl Fisher

BADEL, Agnes, Dr.

Private address:
rue P. Brumier, 46
69300 Calnire

Domains of application:
Quality control of raw materials and finished products; development of products

Applied methods:
IR, HPLC

BERMOND, Dr.
Institut National Agronomique
rue Claude Bernard, 16
75231 Paris
Phone: +33 1 44081646
Fax: +33 1 44081653
E-Mail: bermond@inapg.inra.fr

Domains of application:
Speciation/reactivity of trace elements in soil; oxydation of organic matter (ozonation, fentos rachol, electrochemical oxidation)

Applied methods:
Electrochemical methods; atomic absorption spectrometry; HPLC with various detectors (MS)

BERTHOD, Alain, Dr.
Laboratoire des Sciences Analytiques
Université de Lyon
Bat 308 - D, CPE Lyon
69622 Villeurbanne Cedex
Phone: +33 4 72431434
Fax: +33 4 72431078
E-Mail: berthod@univ-lyon1.fr

Domains of application:
Analytical sciences, analytical chemistry; spectroscopic methods; separation methods; environmental analysis

Applied methods:
HPLC; gas chromatography; countercurrent chromatography; process in separation; spectroscopy; inductively coupled plasma; mass spectrometry

Membership:
ACS; French Chemical Society; Lyon's Club of Chromatography

BUTRUILLE, Jean-Rene, Dr.
Pechiney
Centr'Alp BP 27
38340 Voreppe
Phone: +33 4 76578318
Fax: +33 4 76578459
E-Mail: butruille@crv.pechiney.fr

Private address:
rue Jean Maee, 17
38000 Grenoble

Domains of application:
Industrial research on aluminium and packaging

Applied methods:
Surface analysis; electrochemistry

Membership:
Societé Francaise de Chimie; Electrochemical Society

Awards:
AIPEA 93 (Adcloide-Australia)

CAMEL, Valerie, Dr.
Institut National Agronomique Paris-Grignon
rue Claude Bernard, 16
75231 Paris
Phone: +33 1 44081725
Fax: +33 1 44081653

Domains of application:
Water pollution and water treatment: oxidation of pesticides on other organic compounds using ozonization on electrochemistry, determination of organic pollutants in water and in sludges

Applied methods:
Chromatography (HPLC, GC); solid-phase extraction; mass spectrometry; electrochemistry

Membership:
International Ozone Association; Groupe Francaise de l'International Humic Substances Society; International Association on Water Quality

---

CAVEDON, Jean-Marc, Dr.
CEA
DPE/SPEA-CEA Saclay
91191 Crif-sum-Yvette
Phone: +33 1 63094728
Fax: +33 1 69085254
E-Mail: cavedon@vandaise.cea.fr
Domains of application:
Nuclear energy; nuclear full cycle; isotope separation; analytical chemistry

---

CHEMLA, Marius, Dr., Eng.
Laboratoire d'Electrochimie
Université Pierre et Marie Curie-Paris VI
Place Jussieu, 4
75257 Paris Cedex 95
Phone: +33 1 44273117
Fax: +33 1 44273834
E-Mail:
mchemla@moka.ccr.jussieu.fr

Private address:
Boulevard Arago, 25
75013 Paris

Domains of application:
Electrochemistry; transport in electrolytes; electrochemical and radiochemical analysis

Applied methods:
Interfacial electrochemistry; corrosion; metals and semiconductors

Membership:
Societé Francaise de Chimie; International Society of

Electrochemistry; American Electrochemical Society

Awards:
Gran Prix Societe Francaise de Chimie; Laureat Academie des Sciences Paris

---

COTE, Bertrand, Dr.
SIAP
Boulevard de l'Industrie, BP 8
33530 Bassens
Phone: +33 5 57776550
Fax: +33 5 57776555

Private address:
rue Cazenare, 16
33100 Bordeaux

Domains of application:
Waste treatment (industrial/hazardous waste)
Applied methods:
Incineration/biological treatment; physico-chemical treatment

Membership:
French Chemical Society

DEVILLIERS, Didier, Dr.
Laboratoire d'Electrochimie
Université Pierre et Marie Curie-Paris VI
Place Jussieu, 4
75252 Paris Cedex 05
Phone: +33 1 44273677
Fax: +33 1 44273834
E-Mail: devill@ccr.jussieu.fr

Domains of application:
Electrochemistry: fundamentals and applications; electroanalysis (trace analysis); electrolytes; industrial electrochemistry (electrolysis-batteries)

Applied methods:
Pulse techniques; specific electrodes; impedance spectroscopy; scanning tunnelling microscopy; atomic force microscopy

Membership:
Societé Francaise de Chimie; ISE; IUPAC; ACS

Awards:
French Chemical Society (Division of Analytical Chemistry)

DHAINAUT, David,
rue Desire Linard, 73
08190 Saint Germainmont

Domains of application:
Dorage pH metrique d'une solution de soude carbonatee; dorages simultanes du calcium et du magnesium par complexometrie et gravimetrie; application a la determination de la durete d'une zav; resines extrangeuses e'ions; separation d'un melonge d'ions

Applied methods:
Determination de la concentration des ions sodium et des ions fluorure dans l'eau; analyse qualitative minerale; application a l'identification d'un corps pur; dosage de l'oxygene dissous et determination de la DCO d'une eau

Awards:
U.V. obtenu a 621100

DUCAUZE, Christian J., Chem. Eng, Dr. Sci. Phys.
Institut National Agronomique
Laboratoire de Chimie Analytique
rue Claude-Bernard, 16
75005 Paris
Phone: +33 1 44081644
Fax: +33 1 44081653
E-Mail: ducauze@inapg.inra.fr

Domains of application:
Food quality control; environment

Applied methods:
Chemometrics; TD-NMR; GC/MS; HPLC/MS; fluorimetry; microwave sample puration

Membership:
FECS-DAC

DUCLAIROIR, Christian, Chem. Eng.
Societé Francaise HOECHST
BP 1
60350 Cuise la Motte

DURIN, Leila, Dr.
CEMAGREF
Barrier and Diairage Engineering Research Division
Parc de Tourvaoie
92163 Antony Cedex
Phone: +33 1 40966045
Fax: +33 1 40966270

Domains of application:
Water and environmental engineering department: water resources, waste management and research, natural risks; engineering of water or pollutant/material/hydraulic infrastructre interactions

Applied methods:
Laboratory tests (physical and chemical tests on materials); UV-visible spectrophotometry; micro-FTIR spectrometry

Membership:
Comité Francaise de Geoxynthetiques; Societé Francaise de Chimie

---

EL MURR, Nabil, Dr.
CNRS
University of Nantes
rue de la Houssiniere, 2
44322 Nantes Cedex 2
Phone: +33 2 40741310
Fax: +33 2 40741310
E-Mail: elmurr@chimbio.univ-nantes.fr

Domains of application:
Electrochemistry; bioelectrochemistry; biosensors

Applied methods:
Electrochemical techniques; spectroscopic

FABRE, Paul-Louis, Dr., Eng.
IUT Chimie
Université Paul Sabatier
Av. G. Pompidou
81100 Castres
Phone: +33 5 63621153
Fax: +33 5 63356388

Private address:
rue de la Balance, 39
31000 Toulouse

Domains of application:
Chimie inorganique; electrochimie analytique et appliquee

Applied methods:
Methodes electrochimiques d'analyse; methodes chromatographiques

Awards:
Prix Eugene Schneller, 1980; Prix Division Chimie Analytique Societe Francaise de Chimie, 1984

FARRAZIN, Jean, Dr.
Ecole Nationale Superieure de Chimie de Montpellier
rue de l'Ecole Normale, 8
34296 Montpellier Cedex 5
Phone: +33 4 67144380
Fax: +33 4 67147236
E-Mail: sorrazin 2cit.enscm.fr

Domains of application:
Training of chemical engineers; research in: membrane materials and processes, catalytic materials and catalysis in organic chemistry, macromolecular and molecular heterochemistry

Applied methods:
Electrochemical methods of analysis; AAS; chromatography (GC, GC-MS, HPLC, IEC, LC-MS); election microscopy; thermal analysis (TGA; DSC); adsorption/desorption; spectroscopy (UN, VIS, IR, NMR)

Membership:
Societé Francaise de Chimie; Union des Physiciens; European Society of Membrane Science and Technology; ISE

FOURNIER, Josette, Prof., Dr.
Centre Regional d'Etude del Produits Agropharmaceutique
Université d'Angers
rue Beequorol
41070 Beaucouze

GAL, Jean-Francois, Dr. Sci.
Laboratoire de Chimie Physique Organique
Université de Nice-Sophia Antipolis
Parc Velrose
06108 Nice Cedex 2
Phone: +33 4 92076110
Fax: +33 4 92076111
E-Mail: gal@unice.fr

Private address:
Av. De Pessicart, 183G
06100 Nice

Domains of application:
Mass spectroscopy: ionization techniques, ion/molecule reactions, collisional activation, isotopes, structure and thermochemistry of ions in the gas phase

Applied methods:
Fourier transform ion cyclotron; resonance mass spectrometry (FT-ICR, MS); electrospray ionization; isotope ratio; MS

Membership:
ACS; ASMS; IUPAC; Societe Fracaise de Chimie; Societe Francaise de Spectrometrie de Masse

Awards:
Prix Edouard Calvet (for contribution in the field of calorimetry)

GOMEZ, Francis, Dr.
Rhone-Poulenc Agro
rue P. Baicet, 14-20
69234 Lyon Cedex 09
Phone: +33 4 72852561
Fax: +33 4 72852052
E-Mail: francis.gomez@rp-agro-rp.fr

Domains of application:
Agrochemicals

Applied methods:
All for organic analytical chemistry

GOUPY, Jacques, Dr.
rue Mignet, 1
75016 Paris

Domains of application:
Experimental designs; applied statistics in chemistry

Membership:
Societé Francaise de Chimie; Président de la Division de Chimie Analytique

---

GY, Pierre, Dr.
Av. Jean de Noailles, 14

06400 Cannes
Phone: +33 4 93990983
Fax: +33 4 93990983
E-Mail: gy@pierregy.com

Domains of application:
Echantiollonnage; sampling

HANERY, Florence, Eng.
CREZED
Zone Portuaire
Av. Dreygons Ducas, 291
75520 Liray
Phone: +33 1 30929898
Fax: +33 1 34786797

Domains of application:
Environment

Applied methods:
Chimie analytique: ICP-AAS-FX

HORVATH, Stephane, Eng.
Technologie Servier
BP 1749
45007 Orleans Cedex 1

IRELAND-RIPERT, Jayne, Dr.
Centre Informatique sur la Qualité des Aliments
rue Claude Bernard, 16
75005 Paris
Phone: +33 1 44081662
Fax: +331 44081660
E-Mail: ciqual@inapg.inra.fr

Domains of application:
National Nutrient Databank

JACQUET, Geraldine, Agron. Eng.
Institut National Agronomique Paris-Grignon
Laboratoire de Chimie Analytique
rue Claude Bernard, 16
75005 Paris
Phone: +33 1 44081659
Fax: +33 1 44081653
E-Mail:
jacquet@cardere.grignon.inra.fr

Private address:
Folleville
78850 Thiverval Grignon

Domains of application:
Environment; agroalimentaire

Applied methods:
Atmospheric-pressure ionization mass spectrometry; HPLC-API/MS analysis of organic species and biomolecules

---

JEQUIER, William, Dr.
ELF Atochem
Bd. de la Milliere, 123
13367 Marseille Cedex 11
Phone: +33 4 91241060
Fax: +33 4 91241005

Domains of application:
Analyses poor fabrication; domaine de competence: analyses d'hoile de ricin

Applied methods:
GPC; GPC-SM; HPLC

KAPPENSTEIN, Charles, Dr.
Laboratoire de Chimie Minerale
Faculté des Sciences
Avenue du Rectaur Pineau
86022 Poitiers
Phone: +33 5 49453860
Fax: +33 5 49454020
E-Mail:
charles.kappenstein@cri.univ-poitiers.fr

Private address:
rue de la Bretonnerie, 4
86000 Poitiers

Domains of application:
Characterization of solid catalysts; x-

ray diffraction; electronmicroscopy with EDX; ATG-DTG; atomic absorption spectroscopy

Applied methods:
XPS (ESCA); BET area; chemisorption on metallic surface

Membership:
IUPAC

LABBE, Jean-Pierre, Dr.
Ecole Nationale Superieure de Chimie de Paris
rue Pierre-et-Marie Curie, 11
75231 Paris Cedex 05
Phone: +33 1 44276721
Fax: +33 1 43263738

Private address:
rue Pierre Beregovoy, 62
92110 Clichy

Domains of application:
Corrosion and corrosion inhibitors; analysis (qualitative-quantitative) of products in surfaces (metals-oxides): adsorption, corrosion, corrosion inhibitors, scales, surface cleanliness

Applied methods:
General analytical methods (microgravimetry, titrations, atomic absorption, x-ray diffraction); electrochemical methods; infrared absorption spectrometry (micropollet techniques for quantitative analysis)

MARTELET, Claude, Dr.
Ecole Centrale de Lyon
Ingenierie et Fonationnalisation des Surfaces
BP 163
69131 Ecully Cedex

MONCHATY, Anne, Dr.
Courvoisier S.A.
Place de Chateau, 2
16200 Jarnac
Phone: +33 5 45355651
Fax: +33 5 45355500

Domains of application:
Agroalimentaire (negociant en spiritueux)

Applied methods:
Analyse chromatographique; HPLC; GC (FID, ECD, SCD); GD-AS

NGUYEN, Qui Dad, Dr.
Ecole Centrale Paris
Grande Vrie des Vigres
92295 Chatenay-Malabry
Phone: +33 1 41131105
Fax: +33 1 41131502
E-Mail: cgdas@pem.ecp.fr

Private address:
Allee de la Sacletterie, 1
91190 Gy/Ivette

Domains of application:
X-ray diffraction; x-ray fluorescence; vibrational spectroscopy (IR and Raman spectroscopy); UV-Vis spectrometry

Applied methods:
Remote Raman spectrometry; industrial control by Raman spectrometry

Membership:
American Association for the Advancemente of Science; Societe Francaise de Chimie

Awards:
1st Juize of the French Chemical Society, 1975; NATO Award

NIGRETTO, Jean Maxime, Dr.
Université de Cergy Pontoise
LECMA. Neuville/oise
95031 Cergy Pontoise

Phone: +33 1 34257049
Fax: +33 1 34257071
E-Mail: migretto@paris.u-cergy.fr
Domains of application:
Analytical chemistry; electrochemistry

Membership:
ISE; SFC

PELLERIN, Fernand J., Dr.
Faculté de Pharmacie
Université Paris
92296 Chalenay Malabry

Private address:
rue Desnouettes, 51
75015 Paris

Domains of application:
Organic analytical chemistry; analysis of drugs and plastic material in pharmaceutical and food use

Applied methods:
Spectrometry; chromatography (HPLC, GLC)

Membership:
Academie Nationale de Pharmacie; European Federation of Pharmaceutical Science; Quality Group of IPEC; IUPAC

PICARD, Robert, Chem. Eng.
Compagne Generale des Combustibles Nucleaires (COGEMA)
Phone: +33 4 68320207
Fax: +33 4 68324645

Private address:
Domaine du Grand Robert Route de Gruissan
1100 Narbonne

Domains of application:
Chimie de l'uranium; mesures destintives et non destinctives des tans d'enrichement isotopiques; dosages "complables" et d'exferlise de l'uranium content; dosage de traces, microtraces et submicrotraces d'elements indisurables

Applied methods:
Chemie conventionalle; chromatographie, AA-plasma; "polumetrie ponderales" (methode originales); spectrometries d'emission d'absorption et de masse; mesures isotopiques non destinctives far discrimination neutronique

Membership:
Societé Chimique on France

PITHON, Francis, Eng.
Radiometer Analytical
rue d'Alsace, 72
69627 Villeurbanne

Membership:
ISE

RAFFI, Jacques, Eng.
Research Laboratory of Food Quality
University of Aix-Marseille III
Larqua, D51
13347 Marseille Cedex 20
Phone: +33 4 91985055
Fax: +33 4 91288858
E-Mail: j.raffi@larqua.u-3mrs.fr

Private address:
Av. De l'Armee d'Afrique, 6
13100 Aix en Provence

Domains of application:
Food irradiation; analytical chemistry; gas phase chromatography; electron spin resonance; thermoluminescence

Applied methods:
Methods of identification of

irradiated foodstuffs

Membership:
New York Academy Sciences; French Society of Chemistry; French Society of Analytical Chemistry Experts

---

RAULIN, Francois, Dr., Eng.
Laborotaire Interuniversitaire des Systemes Atmospheriques
Universités Paris 7 et Paris 12
Av. Du General de Gaulle, 61
94000 Creteil Cedex
Phone: +33 1 45171550
Fax: +33 1 45171564
E-Mail: raulin@univ-paris12.fr

Private address:
rue Bourgon, 2
75013 Paris
France

Membership:
Societé Francaise de Chimie; Societé Francaise des Specialistes d'Astronomie; American Geophysical Union; European Geophysical Society; American Astronomy Society

---

ROTH, Etienne, Dr.
Cen Sallay
rue Brancas, 103
92310 Sevres
Phone: +33 1 69089880
Fax: +33 1 69088707

Private address:
etienneroth@compuserve.com

Domains of application:
Nuclear techniques; isotope separation and analysis

Applied methods:
Infrared analysis; cuclear methods of analysis; stoke isotope and isotope detection analysis

Membership:
IUPAC; FECS

RUTLEDGE, Douglas Neil, Bach. Sci.
Institut National Agronomique
rue Claude Bernard, 16
75005 Paris
Phone: +33 1 44081648
Fax: +33 1 44081653
E-Mail: rutledge@inapg.inra.fr

Domains of application:
Food properties

Applied methods:
Time domain cuclear magnetic resmance; chemometrics

Membership:
IUPAC; Societe Francaise de Chimie; FECS-DAC

SABOT, Jean-Francoise, Dr.
Institut des Sciences Pharmaceutiques et Biologiques de Lyon
Faculté de Pharmacie
Av. Rockefellet, 8
69373 Leyon Cedex 08
Phone: +33 1 777000
Fax: +33 4 777156
E-Mail: sabot@univ-lyon1.fr

Domains of application:
Gas chromatography; mass spectrometry associated with stable isotope dilutions; analysis of calculi (urinary, biliary, salivary) by Fourier transform infrared spectroscopy; creation of FT-IR spectral libraries

Applied methods:
GC-mass spectrometry and stable isotope dilutions; molecular modeling (mol. Mechanics, quantum chemistry); separation methods; medical biology; radioimmunology; molecular biology; porphyrins

analysis; analysis of calculi

SELLIER, Nicole, Dr.
Ecole Nationale Superieure de Chimie de Paris
Laboratoire de Spectrometric de Masse
rue Pierre et Marie Curie, 11
75231 Paris Cedex 05
Phone: +33 1 43297276
Fax: +33 1 43257975
E-Mail: sellier@ext.jussieu.fr

Domains of application:
GC/MS/EJ; GC/MS/PCI; GC/MS/NCI of essential oils; determination of chirality using dimethyl ether as reactant gas; differenciation of polycyclic aromatic hydrocarbon (PAHs= by using DME; Reaction ion-molecule

Applied methods:
Mass spectrometry (MS); gas chromatography (GC); GC/MS

THEVENOT, Daniel, Dr.
Laboratoire de Bioelectrochimie et Analyse du Molsen (LABAM)
Faculté des Sciences
Av. Du General de Gaulle, 61
94010 Creteil Cedex
Phone: +33 1 45171625
Fax: +33 1 45171627
E-Mail: thevenot@univ-paris12.fr

Private address:
rue Dareau, 32
75014 Paris
France

Membership:
IUPAC; ISE; International Association on Water Quality

UCHIMURA, Masayuki,
Centre National de la Recherche Scientifique UNM 9987
Laboratoire des Materiaux et Procedes Membranaires
Route de Mende
34033 Montpellier
Phone: +33 4 67613414
Fax: +33 4 67042820
E-Mail: uhimura@gbsa.arpb.univ-momp2.fr

Private address:
Le Surville, esc. 17
34080 Montpellier

Domains of application:
Environment, latte contre la pollution

Applied methods:
Developpement de panneaux a membrane pour la destruction de la canlerpa raxifolid

Membership:
Societé Franco-Japonaise d'Oceanographie (SFJO)

ZHANG, Ben-Li, Dr.
Laboratoire d'Analyse Isotopique et Electrochimique de Metabolismes
CNRS Upres-A 6006
rue de la Houssiniere, 2
92208 Nantes
Phone: +33 2 40373136
Fax: +33 2 40373169
E-Mail: zhang@chimbio.univ-nantes.fr

Domains of application:
Study of isotopic effects and mechanisms of chemical and biological reactions; development of methods for identification of products

Applied methods:
Analysis by isotopic ratio NMR and isotopic ratio MS

ALBERT, Klaus, Prof., Dr.
Institut für Organische Chemie
University of Tübingen
Auf der Morgenstelle 18
72076 Tübingen
Phone: +49 7071 2975335
Fax: +49 7071 295875
E-Mail: klaus.albert@uni-tuebingen.de

Private address:
Hainbuchenweg 25
72076 Tübingen

Domains of application:
Isolation and structure elucidation of phytopharmaceuticals; functional food analysis; development of tailored stationary phases for chromatographic applications; investigation of molecular recognition processes by NMR techniques

Applied methods:
HPLC-NMR coupling; SFC-NMR coupling; HPLC-MS coupling; solid-state NMR; high-resolution magic angle spinning NMR; CEC-NMR coupling

Membership:
GDCh; ACS

BAHADIR, Mufit, Prof., Dr.
Institute of Ecological Chemistry and Waste Analysis
Technical University of Braunschweig
Hagenring 30
38106 Braunschweig
Phone: +49 531 3915960
Fax: +49 531 3915799
E-Mail: m.bahadir@tu-bs.de

Domains of application:
Environmental chemistry; pesticide chemistry; ecotoxicology; combustion process; dioxins and furans; organic residue analysis; inorganic analysis; hazardous waste chemistry; waste mining; recycling processes

Applied methods:
AAS (GF, H, HG, CV); ICP-OES; IC; HPLC; GPC; GC; GC/MS; 14C-techniques; FT-IR; $^{1}H/^{13}C$-NMR; group parameters (TOC, TOX, JOD, COD, etc.)

Membership:
MESAEP: Vice-President

BERNDT, Harald, Prof., Dr.
Institut für Spektrochemie und Angewandte Spektroskopie
Bunsen-Kirchhoff Str. 11
44139 Dortmund
Phone: +49 231 1392180
Fax: +49 231 1392480
E-Mail: berndt@isas-dortmund.de

Domains of application:
Elemental trace analysis; elemental speciation; sample introduction in atomic spectrometry

Applied methods:
All types of atomic absorption spectrometry; ICP/OES; HPLC

Awards:
Prof. Visitante, Universidad de Concepción, Chile

BROEKAERT, José A.C., Prof., Dr.
Department of Analytical Chemistry
University of Leipzig
Linné-Strasse 3
04103 Leipzig
Phone: +49 341 9736100
Fax: +49 341 9736115

E-Mail:
broekaer@sonne.tachemie.uni-leipzig.de

Private address:
Kleiberweg 14
44269 Dortmund

Domains of application:
Analytical chemistry with special reference to plasma atomic spectrometry and electrochemistry with special reference to environmental applications, life sciences and avoiding materials

Applied methods:
Atomic spectrometry (optical environ, atomic absorption, mass spectrometry); electrochemistry; gas and liquid chromatography

Membership:
AAAS; SAS

Awards:
Kasimir Pulaski Medal, 1994

CAMMANN, Karl, Prof., Dr.
Institut für Chemo- und Biosensorik
University of Münster
Mendelstr. 7
48149 Münster
Phone: +49 251 9802800
Fax: +49 251 9802802
E-Mail: cammann@uni-muenstev.de

Domains of application:
Environmental analysis; trace analysis; biochemical analysis

Applied methods:
Spectroscopy; chromatography; electroanalysis; chemical and bionsensors

Membership:
ACS; ISE; I.U.P.A.C.

Awards:
Oce-van-der-Grinten Environmental Award 1967; Technology Transfer Award Ministry of Science 1987 and Education; Inventor Ward of the State North-Rhine-Wetsphalia 1997 Award. Pregl-medal of the Austrian Society for Analytical Chemistry 1979.

DANZER, Klaus, Prof., Dr. rer.nat. habil.
Institute of Inorganic and Analytical Chemistry
Friedrich-Schiller-University Jena
Lessingstrasse 8
07743 Jena
Phone: +49 3641 635925
Fax: +49 3641 635930
E-Mail: klaus.danzer@rz.uni-jena.de

Private address:
Am Friedensberg 4
07745 Jena

Domains of application:
Trace analysis of inorganic materials; organic trace analysis; non-invasive determination of blood glucose; chemometrics: multivariate data analysis, neutral networks, pattern recognition; trace components in wine

Applied methods:
Analytical atomic spectroscopy (ICP, arc, spark, solid sampling AAS); solid phase micro extraction; ICP-MS; GC; HPLC; IR- and NIR spectroscopy

Membership:
IUPAC: Titular member of Commission V.1: General Aspects of Analytical Chemistry; ICS; New York Academy of Sciences

---

DUNEMANN, Lothar, Prof., Dr.
Medical Institute of Environmental

Hygiene
PO Box 10 37 51
40028 Düsseldorf
Phone: +49 211 3389375
Fax: +49 211 3389331
E-Mail: dunemann@uni-duesseldorf.de

Domains of application:
Environmental monitoring; human biomonitoring; trace analysis of heavy metals and platinum group metals and volatile organic compounds; trace analysis of persistent chemicals and metabolites

Applied methods:
AAS, ICP-MS, HR-ICP-MS, GC (FID, ECD, PID), GC-MS, GC-PFPD, HPLC (UV, fluorescence, MS, ED), CE

Membership:
GDCh, Deutscher Hochschulverband; GMS

Awards:
Award of the section "Analytical Chemistry" of the GDCh

EBEL, Siegfried, Prof., Dr.
Department of Pharmacy
Am Hubland
97074 Würzburg
Phone: +49 931 8885460

Domains of application:
Instrumental analysis; drug analysis; analysis of drug stability

Applied methods:
TLC; HPLC; UV; IR; NIR

Membership:
GDCh; BPCh; DPhCh

EINAX, Jürgen W., Prof., Dr. rer.nat. habil.
Institute of Inorganic and Analytical Chemistry
Chair of Environmental Analytical Chemistry
Friedrich Schiller Universität
Lessingstrasse, 8
07743 Jena
Phone: +49 3641 948190
Fax: +49 3641 948172
E-Mail: juergen.einax@uni-jena.de

Domains of application:
Environmental analysis; chemometrics; pollution assessment; sampling strategies; speciation

Applied methods:
Atomic spectroscopy; electroanalysis; liquid and gas chromatography

---

EMONS, Hendrik, Dr. Sci., Dr. rer.nat.
Institute of Applied Physical Chemistry
Research Center of Jülich
52425 Jülich
Phone: +49 2461 612914
Fax: +49 2461 612493
E-Mail: h.emons@fz-juelich.de

Domains of application:
Environmental research with respect to terrestrial and aquatic ecosystems (environmental analysis, environmental specimen banking, interfacial structures and processes in the environment, bioindication, processes of water and soil treatment)

Applied methods:
Atomic spectroscopy (electrothermal AAS, cold vapor AAS, hydride generation AAS, atomic emission spectroscopy); electroanalysis (voltammetry, potentiometry, amperometry); mass spectroscopy (ICP-AS, IDMS with thermal ionization); HPLC with DAD, PAD

Germany

Membership:
ISE; European Society of Electroanalytical Chemistry: Chairman of ESEAC 2000; Advisory Board of "The Analyst" and "Analytical Communications"

Awards:
First Prize at the "Competition of Young Chemists", 1982; Humboldt Fellowship 1991/92

ENGELHARDT, Heinz, Prof., Dr.
Instrumentelle Analytik/Unweltanalytik
Universität des Saarlandes
Im Stadtwald
66123 Saarbrücken
Phone: +49 681 3023293
Fax: +49 681 3022963
E-Mail: iaua@rz.uni-sb.de

Private address:
Kiefernstr. 23
66129 Saarbrücken

Domains of application:
Synthesis of stationary phases; problem solving for pharmaceutical, chemical and environmental analysis; development of new separation systems; sample preparation techniques

Applied methods:
Liquid, gas and supercritical fluid chromatography; capillary electrophoresis; capillary electrochromatography

Membership:
GDCh; American Chromatographic Society; Board of Subdivision of Analytical Chemistry

Awards:
1992: Dal Nogare Award; 1993: Martin-Medal (Chromatographic Society of Great Britain)

ENGEWALD, Werner, Dr. rer.nat.habil.
Institute of Analytical Chemistry
University of Leipzig
Linnestr. 3
04103 Leipzig
Phone: +49 341 9736061
Fax: +49 341 9736115
E-Mail: engewald@rz.uni-leipzig.de

Private address:
Kriekauer Str. 13
04425 Taucha

Domains of application:
Chromatographic multicomponent and trace analysis of organic compunds/mixtures; application of chromatographic methods in environmental analysis

Applied methods:
GC, including GC/MS, GC/AED, pyrolysis-GC/MS, adsorptive enrichment, thermal desorption; HPLC and API

Membership:
Member of the Chromatographic Society

Awards:
Friedrich-Wöhler-Preis der ehemaligen Chemischen Gesellschaft der DDR

EPPERT, Gunter J., Dr. habil.
Sepserv
Separation Service Berlin
Helmholtzstr. 2-9
10587 Berlin
Phone: +49 30 3930991
Fax: +49 30 3934925

Domains of application:
Synthesis and investigation of new chromatographic packings

Applied methods:
Chromatography (HPLC)

Membership:
GDCh

FABRY, Laszlo Georg, Dr. rer.nat. Univ.
Wacker Siltronic AG
PO Box 1140
84479 Burghausen
Phone: +49 8677 834860
Fax: +49 8677 835954
E-Mail: l.fabry@wacker.de

Domains of application:
Ultratrace analysis of elements relevant to semiconductor quality of pure silicon, in products, process media and environment, GFAAS, ICP-OES, ICP-MS, TXRF, INAA, IC, CE, GFA, DLTS, PCD, HgCV

Applied methods:
LT-FTIR, FTIR, PL

Membership:
American Scientific Affiliation; The New York Academy of Sciences; The Institute of Electric and Electronics Engineers; GDCh; Semiconductor Equipment and Materials International

Awards:
Candidate of Chemical Societies, Hung. Acad. of Sci. 1981

FRIEBOLIN, Horst, Prof., Dr.
Organisch-Chemisches Institut of University of Heidelberg
Im Neuenheimer Feld 270
69120 Heidelberg
Phone: +49 6221 548472
E-Mail: jqg@ix.urz.uni-Heidelberg.de

Private address:
Max-Reger-Str. 32
69121 Heidelberg

Domains of application:
NMR

Applied methods:
NMR

FRIMMEL, Fritz Hartmann, Prof., Dr.
Department of Chemical Engineering
Institute of Water Chemistry
Richard-Willstätter-Allee 5
76131 Karlsruhe
Phone: +49 721 6082580
Fax: +49 721 699154
E-Mail: fritz.frimmel@ciw.uni-karlsruhe.de

Domains of application:
Analysis of natural water components and micropollutants; speciation; isotopic ratios; characterization of natural organic matter (humic substances, refractory organic acids); analysis of wastewater; sediments; oil

Applied methods:
HPLC-DAD/fluorescence/MS; HPLC-scintillation detection; IC; IC-pulsed amperometric detection; LC-organic carbon detection/UV/fluorescence; GC-ECD/FID; GC-ion trap MS; purge-and-trap GC; headspace GC; GC-AED; capillary electrophoresis-DAD; F-AAS; GF-AAS

Membership:
American Water Works Association; International Association on Water Quality; ACS; International Humic Substances Society

Awards:
Preis des Nachwuchses auf dem Gebiet der Wasserchemie, Fachgruppe Wasserchemie in GDCh, 1975; Paulaner Forschungspreis, 1984; Environmental Education

Germany

Award SETAC-Europe/ABC Laboratories, 1992

GAUGLITZ, Gunter, Prof., Dr.
Institute of Physic and Theoretical Chemistry
Eberhard-Karls-Univ. Tübingen
Auf der Morgenstelle 8
72076 Tübingen
Phone: +49 7071 297692
Fax: +49 7071 296910
E-Mail: guenter.gauglitz@ipc.uni-tuebingen.de

Private address:
Panoramastr. 54
72070 Tübingen

Domains of application:
Spectroscopy; ultraviolet; visible and near infrared spectroscopy; surface analysis; fluorescence, chemical and biochemical sensing

Applied methods:
Absorption; reflectance; luminescence; microfluidic; multicomponent analysis; chemometrics; optical transducers; spectroscopy; spectroscopy at interfaces and of thin films; ellipsometry; surface plasmon resonance

Membership:
GDCh; IUPAC

GOLLOCH, Alfred, Prof., Dr. rer.nat.
Fachgebiet Instrumentelle Analytik
Gerhard-Mercator-University of Duisburg
Lotharstrasse 1
47047 Duisburg
Phone: +49 203 3793311
Fax: +49 203 3792108
E-Mail: golloch@lims.uni-duisburg.de

Private address:
Schönauer Bach 21
52072 Aachen

Domains of application:
Fundamentals and application of atomic absorption and emission spectrometry; fundamentals and application of molecular spectroscopy, infrares spectrometry; thermal analysis; chemometrics; materials analysis

Applied methods:
AAS; ICP atomic emission spectrometry; MIP atomic emission spectrometry; sliding spark spectrometry; differential thermal analysis; differential scanning calorimetry; ion mobility spectrometry; total reflection x-ray

Awards:
Borchers-Plakette RWTH Aachen

GROTE, Manfred, Prof., Dr.
Department of Analytical Chemistry
University of Paderborn
Warburgerstrasse 100
33098 Paderborn
Phone: +49 5251 60 2191
Fax: +49 5251 603549
E-Mail: mg@chemie.uni-paderborn.de

Domains of application:
Method development and application: element species analysis (phytoavailability of toxic metals), determination of pharmaceuticals in the environment; biochemical-clinical analysis: determination of antioxidants in bodyfluids; development of optical sensors

Applied methods:
DCP-OES; AAS, ion chromatography; HPLC (UV-DAD, EC, fluorescence-detection); GC (ECD, FID); GC-MS; UV-, FT-IR-

spectroscopy

Membership:
GDCh: Fachgruppe Analytische Chemie; Geschichte der Chemie; MESAEP (Mediterranean Scientific Association of Environmental Protection)

GRÜNDLER, Peter, Dr. rer.nat.
Chemical Department
Rostock University
Buchbinderstrasse 9
18051 Rostock
Phone: +49 381 4981754
Fax: +49 381 4981763
E-Mail:
peter.gruendler@chemie.uni-rostock.de

Private address:
Gartenstrasse 12
18119 Rostock

Domains of application:
Electrochem. sensor with heated microelectrodes; environmental technology; anal. sensor for H_2S; anal. det. of organic in sea water or soils

Applied methods:
Electrochemistry; chromatography (GC, GC-MS, HPLC); atomic absorption spectrometry

---

GÜTLICH, Philipp, Prof., Dr.
Institut für Anorganische Chemie und Analytische Chemie
Johannes Gutenberg-Universität Mainz
Staudinger Weg 9
55099 Mainz
Phone: +49 6131 392373
Fax: +49 6131 392990
E-Mail: p.guetlich@uni-mainz.de

Private address:
Georg-Büchner-Str. 9
64380 Rossdorf

Membership:
GDCh; RSC; ACS; BPCh; DGM; DPG

HÄGELE, Gerhard, Prof., Dr.
Institut für Anorganische Chemie und Strukturchemie I
Heinrich-Heine-University of Düsseldorf
Universitäatsstr. 1
40225 Düsseldorf
Phone: +49 211 8113140
Fax: +49 211 8113085
E-Mail: haegele@uni-dusseldorf.de

Private address:
Kerschensteiner Weg 13, 40723 Hilden

Domains of application:
Inorganic chemistry; analytical chemistry; computer assisted titration; hard- and software development; hyphenated techniques; NMR-, UV/VIS-controlled titrations; NMR-spectroscopy

Applied methods:
Potentiometry; NMR; UV/VIS

Membership:
ACS; The Royal Chemical Society

HEDRICH, Martina, Dr.
Laboratory I-43
BAM
Unter den Eichen 87
12205 Berlin
Phone: +49 30 81041142
Fax: +49 30 81041147
E-Mail: martina.hedrich@bam.de

Domains of application:
Determination of element contents; certification analysis of reference materials; gaseous reference

Germany

materials

Applied methods:
Photon activation analysis; neutron activation analysis; gravimetric preparation of primary reference gas mixtures; gas chromatography

Membership:
ISO/TC-183; Convener WG-16 "Fluorine and Chlorine in Copper, bend and Zinc Concentrates"

HEUMANN, K.G., Prof., Dr.
Institute of Inorganic and Analytical Chemistry
Johannes Gutenberg-Universität Mainz
Bacherweg 24, 55099 Mainz
Phone: +49 6131 395882
Fax: +49 6131 393369
E-Mail: heumann@mail.uni-mainz.de

Private address:
Lange Schneise 5
64673 Zuringenberg

Domains of application:
Trace analysis of elements and element species; isotope ratio measurements; applications in high-tech materials and environmental samples

Applied methods:
(isotope dilution) mass spectrometry (ICP-AS and FIAS); AAS; AES; AFS; electroanalytical methods DPASV; HPLC; GC; IC

Membership:
IUPAC

Awards:
Award of the Technical University of Darmstadt, 1986; Oce-van-der-Grinten Award for environmental analytical chemistry, 1987

HIRNER, Alfred, Prof., Dr.
Institute of Environmental Analytical Chemistry
University of Essen
45117 Essen
Phone: +49 201 1833950
Fax: +49 201 1833951
E-Mail: alfred.hirner@uni-essen.de

Domains of application:
Contaminant mobility testing and prognosis; chemical fingerprinting; speciation, metall(oid)organic compounds in the geo-, amthroposphere and medicine

Applied methods:
XRD; XRS; ICP-OES; ICP-MS; GC; HPLC; IC; GC-MS; GC/ICP-MS

Membership:
Geochemical Society

Awards:
Heisenberg fellow (DFG)

HOYER, Georg A., Prof., Dr.
Schering AG
Institute of Physical Chemistry
Müllerstr. 178
13342 Berlin
Phone: +49 30 46814709
Fax: +49 30 46818078

Private address:
Deutschtaler Str. 41, 12355 Berlin

Domains of application:
General physical chemistry; structure analysis; trace analysis; computational chemistry; information technology; computer validation

Applied methods:
All methods used in these areas

Membership:
GDCh

KARAS, Michael, Prof. Dr.
Instrumentelle Analytische Chemie
Johann Wolfgang Goethe University
Theodor-Stern-Kai 7
60590 Frankfurt
Phone: +49 69 63017624
Fax: +49 69 6313984
E-Mail: karas@chemie.uni-frankfurt.de

Domains of application:
Bio mass spectrometry; fundamentals of ion generation and methodic developments for new applications

Applied methods:
Electrospray-ionization (ESI); matrix-assisted lase desorption-ionization; time-of-flight; quadrupo-ion/trap-mass spectrometers

Membership:
DGMS; ASMS; GDCh

Awards:
Mattauch-Herzog Award (AGMS Germany); ASMS Award for Distinguished Contribution to MS

KETTRUP, Antonius, Prof. Dr. rer.nat.
GSF Forschungszentrum für Umwelt und Gesundheit
Institut für Ökologische Chemie
Ingolstädter Landstr. 1
85764 Neuherberg
Phone: +49 89 31874047
Fax: +49 89 31873371
E-Mail: kettrup@gsf.de

Private address:
Rumbecker Höhe 10
59821 Arnsber

Domains of application:
Environmental Analysis and Bioavailability metal speciation, analysis of humic material ecotoxicology and bioassays waste analysis and evaluation pyrolysis of chemicals and waste thermal analysis (REMPI/TOFMS) analysis of indoor/outdoor pollutants, biomon

Applied methods:
GC; GC/MS; HPLC; HPLC/MS-MS; EC; IC; SFC; headspace-GC-AAS; ICP-OEC; ICP-MS; voltammetry; X-Ray; TA; TA/MS; DSC, NMR, REMPI/TOFMS

Membership:
International Society for Ecotoxicology and Environmental Safety (President); International Confederation on Thermal Analysis and Calorimetry; European Society for Thermal Analysis, Calorimetrie; ACS

Awards:
1984: Netzsch-GEFTA-Preis für Thermische Analyse der Gesellschaft für Thermische Analyse und Kalorimetrie; 1995: UNESCO-Chair for Environmental Engineering Chinese Academy of Sciences; 1999: Member of the North-Rhine-Westphalian Academy of Sciences, Düsseldorf

---

KLOCKOW, Dieter, Prof., Dr.
Institut für Spektrochemie und Angewandte Spektroskopie
Bunsen-Kirchhoff-Str. 11
44139 Dortmund
Phone: +49 231 1392103
Fax: +49 231 1392300
E-Mail: klockow@isas-dortmund.de

Private address:
Knappstr. 1a
44267 Dortmund

Domains of application:
Trace analysis, element speciation, atmospheric trace compounds, miniaturized analytical systems,

process analytical chemistry

Applied methods:
Atomic emission and absorption spectrometry, including laser and plasma techniques; MS; NMR-, IR- and Raman spectrometry; x-ray fluorescence spectrometry; microbeam and surface analysis; GC; LC

Membership:
Member of Executive Committee of IAEAC; AEST

Awards:
Award of the Analytical Chemistry Division of the German Chemical Society

---

KNOCHEL, Arnold, Dr. rer.nat., Dipl.-Chem.
Institut für Anorganische und Angewandte Chemie
University of Hamburg
Martin-Luther-King Platz 6
20146 Hamburg
Fax: +49 40 41232893
E-Mail: knochel@chemie.uni-hamburg.de

Domains of application:
Methodic development and applications of ultratrace element analysis for ecological and biological questions; analytical applications of synchrotron radiation

Applied methods:
NAA; electroanalysis; AAS; AES; ICP-MS; TRFA; GC, HPLC; SFC; IC
Membership:
GDCh

KOPPENHOEFER, Bernard, Prof. Dr. rer.nat.
Institut für Organische Chemie
Tübingen
Auf der Morgenstelle 18
72076 Tübingen
Phone: +49 7071 297203
Fax: +49 7071 292037
E-Mail: kippenhoefer@uni-tuebingen.de

Private address:
Eichenweg, 2
72076 Tübingen

Domains of application:
GC and GC-MS of enantiomers, application in biochemistry and clinical chemistry; CE of chiral drugs; database Chirbase/GC, Chirbase/CE, Chirbase/Flavor; applications of chirally modified sensors; investigation and synthesis of chiral polysiloxanes and CDs

Applied methods:
GC and GC-MS; NMR; microcalorimetry; sensors and actuators

Membership:
Member of 10 scientific communities

KORTE, Ernst-Heiner, Dr.
ISAS-Institut für Spektrochemie
Department Berlin
Rudower Chaussee 5
12489 Berlin
Phone: +49 30 63923544
Fax: +49 30 63923544
E-Mail: korte@isas.fta-berlin.de

Domains of application:
Application-oriented basic research for developing and advancing methods of analytical spectroscopy (e.g. Of gaseous and particulate atmospheric componentes, laser ablated biological material, thin metallic, semiconducting, or organic layers

Applied methods:
IR, ellipsometry, GC/IR, ICP-MS and -OES, UV/VIS/NIR echelle spectrometers

Membership:
DAC-FECS (Secretary); Editorial Boards of Fresenius J. Anal. Chem., Vibr. Spectrosc.; Spectrosc. Letters

---

KRIVAN, Viliam, Prof. Dr.
University of Ulm
Albert-Einstein-Allee 11
89081 Ulm
Phone: +49 73163175
Fax: +49 731 60280400
E-Mail: viliam.krivan@chemie.uni-ulm.de

Private address:
Johann-Miller-Strasse 28
89081 Ulm

Domains of application:
Slurry and solid sampling techniques for electrothermal atomic absorption spectrometry and for electrothermal vaporization atomic emission and mass spectrometry; separation method for trace and ultratrace element analysis

Applied methods:
Flame, electrothermal and hydride generation atomic absorption spectrometry; inductively coupled plasma atomic emission and mass spectrometry; x-ray fluorescence spectrometry; neutron-, fest neutron- and charged particle activation analysis.

Membership:
ACS; SAS; GDCh; Slovak Chemical Society

LINSCHEID, Michael W., Prof. Dr.
Institut für Chemie
Humboldt University of Berlin
Hessische Str. 1-2
10115 Berlin
Phone: +49 30 20937575
Fax: +49 30 20936985
E-Mail:
michael.linscheid@chemie.hu-berlin.de

Private address:
Lipaerstr., 12a
12203 Berlin

Domains of application:
Interactions of xenobiotic compounds with DNA; micropollutants; species analysis; structure elucidation of natural compounds

Applied methods:
Hyphenated techniques (LC/MS; CE/MS; LC/ICPMS), FTMS enzymatic methods, micro separation techniques

Membership:
ACS; ASMS; ISSX; GDCh; DGMS; IUPAC (exofficer, now affiliated member)

Awards:
Former chairman of the DGMS.
STA Fellowship (JICST, Japan).

MARX, Günter, Prof. Dr.
Department of Chemistry
Technical University of Chemnitz
Strasse der Nationen 62
09111 Chemnitz
Phone: +49 371 5311713
Fax: +49 371 5311371
E-Mail: guenter.marx@chemie.tu-chemnitz.de

Private address:
Stollberger Str. 27
09119 Chemnitz

Domains of application:
Surface and interface investigations

of thin films and coatings, characterization of advanced materials (ceramics, fibres, composites); trace analysis of Se and Te in biomaterials; environmental analysis

Applied methods:
UV-, VIS- IR-, FT-IR- and Raman spectroscopy; electromicroscopy, X-ray diffraction; electron probe microanalysis; AAS; flow discharge optical emission spectroscopy; MS; GC; LC

MAURER, Wolfgang, Prof. Dr.
Analytische Chemie II
University of Siegen
Adolf-Reichwein-Str. 2
57068 Siegen
Phone: +49 271 7404240
Fax: +45 2744 8440

Private address:
San Fernando 14
57562 Herdorf

Domains of application:
Drinking water quality; soil pollution; composition and effects of cooling lubricants

Applied methods:
AAS; HPLC; IC; CZE

MEUSINGER, Reinhard, Dr.
Institute of Analytical Chemistry
University of Leipzig
Linnestr. 3-5
04103 Leipzig
Phone: +49 341 9736052
Fax: +49 341 9736115
E-Mail: rmeusi@rz.uni-leipzig.de

Domains of application:
Molecular structure determination; development of analytical methods; environmental analytical chemistry

Applied methods:
NMR; QSPR

NAU, Heinz, Prof. Dr.
Department of Food Toxicology
Centre of Food Sciences
Bischofsholer Damm 15
30173 Hannover
Phone: +49 511 8567600
Fax: +49 511 8567680
E-Mail: hnau@lebensmittel-tiho-hannover.de

Domains of application:
Analysis of food contaminants: toxicological relevance; felate metabolism and neural total defects; vitamin A metabolism and teratogenesis; gene expression in cell cultures as toxicological model system

Applied methods:
HPLC; GC; GC-MS; HPLC-MS; PCR; RT-PCR; gene expression; cell culture in vivo teratological model

Membership:
GÖCh; GDCh; AAAS; ASMS; ACS; ISSX; American Academy Toxicology; New York Academy Sciences

Awards:
Dr. h.c., Univ. Uppsala, Sweden, 1991; Warkary Award, Teratology Society, 1992

NEHLS, Irene, Dr. rer.nat.habil.
Federal Institute of Materials Research and Testing
Rudower Chaussee 5
12489 Berlin
Phone: +49 30 63925816
Fax: +49 30 63925972
E-Mail: irene.nehls@bam.de

Private address:
Albert-Wiebach-Str. 14b

14513 Teltow

Domains of application:
Development of new methods for the organic environmental analysis; development, preparation and certification of reference materials of different matrices (soil, sediment, wood, plants, waste oil, etc.) with "natural" contamination (PAH, PCB, PCP, BTEX)

Applied methods:
Gas chromatography with various detections (GC-MS, GC-HRMS, GC-ECD, GC-FID, GC-AED, GC-FTIR); HPLC (HPLC-DAD, HPLC-FD, HPLC-UV, HPLC-IR); AOX, EOX, TOC, CHN, CE

Membership:
Eurachem member of task group Education and Training

Awards:
"Humboldt medal" in silver for university education

NEIDHART, Bernd, Prof. Dr.
Institute of Physical and Chemical Analytics
GKSS Research Center
Max-Planck-Strasse
21502 Geesthacht
Phone: +49 4152 872345
Fax: +49 4152 872366
E-Mail: bernd.neidhart@gkss.de

Domains of application:
Environmental; AQS

Applied methods:
TXRF; ICP; INAA; AAS; GC; LC; BOCA; laser analytical techniques; LIDAR

Membership:
EURACHEM: Chairman of WG "Education and Training"

NIEMAX, Kay, Prof. Dr. hc
Institut für Spektrochemie und Angewandte Spektroskopie
Bunsen-Kirchhoff-Str. 11
44139 Dortmund
Phone: +49 231 1392101
Fax: +49 231 1392310
E-Mail: niemax@isas-dortmund.de

Private address:
Am Wiesenberge 2
58239 Schwerte-Geisecke

Domains of application:
Element speciation; development of instruments and methods; miniaturized analytical systems; microdomain and surface analysis; material analysis and technical processes; organics in water

Applied methods:
Atomic emission and absorption spectrometry, including laser and plasma techniques; inorganic and organic mass spectrometry; NMR-, IR- and Raman spectrometry; x-ray fluorescence spectrometry; microbeam and surface analysis; gas and liquid chromatography

Membership:
GDCh; DPG; SAS; OSA; DBG

NIESSNER, Reinhard, Prof. Dr.
Institute of Hydrochemistry
Technical University of Munich
Marchioninistr. 17
81377 Munich
Phone: +49 89 70957981
Fax: +49 89 70957999
E-Mail: r.niessner@ws.chemie.tu-muenchen.de

Private address:
Alter Berg 10a
82319 Starnberg

Domains of application:
Hydrochemistry; aerosol chemistry;

environmental analytical chemistry; laser spectroscopy; immunoassay development

Applied methods:
Immunoassays; chromatography (GC/MS, HPLC, CZE, HPTLC/AMD); laser spectroscopy (fluorescence, photoacoustics, plasmaemission); total reflection – XRFA

Membership:
American Association for Aerosol Research (membership); Association for Aerosol Research (membership), European Science Foundation (membership).

Awards:
Heinrich-Emanuel-Merck-Prize, 1990; Smoluchowski Aerosol Award, 1991; Prize in Analytical Chemistry of the German Association of Analysis, 1988; Fritz-Pregl-Medal of the Austrian Analytical Chemistry Association, 1996; Fresenius-Award for Analytical Chemi

---

ORTNER, Hugo M., Prof. Dr.
Technische Universität Darmstadt
FG Chemische Analytik
Petersenstr. 23
64287 Darmstadt
Phone: +49 6151 166309
Fax: +49 6151 166378
E-Mail: h.ortner@hrzpub.tu-darmstadt.de

Private address:
Gutenbergstr. 16
64342 Seeheim-Jugenheim

Domains of application:
Materials characterization; particle characterization; environmental analytical chemistry; high temperature corrosion of metals and graphite

Applied methods:
Bulk analytical methods: XRS, TXRF, FAAS, GFAAS, ICP-OES, ion chromatography; topochemical methods: high resolution scanning electron microscopy, electron probe microanalysis, secondary ion mass spectrometry

Membership:
Managing Editor of "International Journal of Refractory Metals and Hard Materials" (Elsevier), since 1991; Advisory Board of Mikrochimica Acta, Fresenius Journal of Analytical Chemistry

Awards:
Fritz Feigl Award of the Austrian Society of Microchemistry and Analytical Chemistry (1970); Richard Kieffer Award of the International Journal of Refractory and Hard Metals, London 1996

OTTO, Mathias, Prof. Dr. rer.nat. habil.
Institute for Analytical Chemistry
Technical University of Mining and Technology
Leipziger Str. 29
09596 Freiberg
Phone: +49 3731 393468
Fax: +49 3731 393666
E-Mail: otto@arsen.anch.tu-freiberg.de

Domains of application:
Elemental and organic analysis for investigating the environment, materials as well as living systems, electrochemical and optical sensors; interpretation systems

Applied methods:
ICP-AES; XRF; HPLC; GC; UV/VIS; IR; NIR; NMR; MS

Membership:
IUPAC Com. V.1; Advisory Board

of Fresenius J. Anal. Chem., Mikrocim Acta (Wien); Chemometrics Int. Lab. Systems

Awards:
Wöhler-Preis, 1984; Fritz-Feigl-Preis, 1990

PERNICKA, Ernst, Prof. Dr.
Max-Planck-Institut für Kernphysik
PO Box 103380
69029 Heidelberg
Phone: +49 6221 516484
Fax: +49 6221 516324
E-Mail: perni@kosmo.mpi-hd.mpg.de

Private address:
Hollmuthstr. 22
69151 Nechasgemind

Domains of application:
Inorganic analysis applied to geology and archeology

Applied methods:
AAS; x-ray fluorescence analysis; neutron activation analysis; ion chromatography

Membership:
GÖCh; GDCh; Deutsche Mineralogische Gesellschaft; Society for Archaeological Science

---

PRZYBYLSKI, Michael, Prof. Dr. rer.nat.
Faculty of Chemistry
University of Konstanz
78457 Konstanz
Phone: +49 7531 882249
Fax: +49 7531 883097
E-Mail: michael.przybylski@uni-konstanz.de

Domains of application:
Biopolymer structure elucidation; protein identification; structure analysis; immuno-analytical application; peptide/protein antigen analysts/molecular recognitions structures; supramolecular interactions of biopolymers; cell membranes

Applied methods:
High-resolution chromatography (HPLC-GPC): high resolution electrophoresis; nanoseparation/chip techniques; biopolymer mass spectrometry (ES, MALDI, LC-MS); CD; NMR spectroscopy; x-ray crystallization; protein chemistry

Membership:
President of DGMS, 1998; member Editorial Board of several international journals

Awards:
St Denis Award in Cancer research, 1982; guest professorship of several European Universities

---

PYELL, Ute, Dipl.-Chem., Dr. rer.nat.
Fachbereich Biologie/Chemie
Universität GH Kassel
Heinrich-Plett-Strasse 40
34109 Kassel
Phone: +49 561 804 4327
Fax: +49 561 804 4480
E-Mail: pyellute@uni-kassel.de

Domains of application:
Environmental analysis; food analysis

Applied methods:
Micellar electrokinetic chromatography (MEKC); capillary electrochromatography (CEC); hydride generation atomic absorption spectrometry (HG-AAS)

RICHTER, Ernst-Ludwig, Prof. Dr.
Staatliche Akademie der Bildenden

Künste
Am Weissenhof 1
70191 Stuttgart
Phone: +49 711 2575261
Fax: +49 711 763252

Domains of application:
Scientific examination of antique silver and gold objects; scientific examination of paintings

Applied methods:
OES (DC-ARC excitation, laser-microprobe); AAS)flame, graphite furnace); XRD; micro-chemistry

Membership:
International Institute for conservation; International Council of Museums

---

SALZER, Reiner, Prof. Dr.
Institute of Analytical Chemistry
Dresden University of Technology
01062 Dresden
Phone: +49 351 4632631
Fax: +49 351 4637188
E-Mail: reiner.salzer@chemie.tu-dresden.de

Domains of application:
IR and Raman spectroscopy. Bioanalytical techniques. Electronic media in university education

Applied methods:
IR-Raman spectroscopy; fluorescence spectroscopy; surface plasmon resonance; spectroscopic imaging; patch-clamp; mass spectroscopy; chromatography; multimedia methods in analytical chemistry

Membership:
Chairman: Div. Analytical Chemistry of the German Chemical Society GDCh

Awards:
Foreign member: Norwegian Academy of Science and Humanities Oslo

---

SCHOLZ, Fritz, Dipl. Chem., Dr. Sc. nat., Prof.
Ernst-Moritz-Arndt-Universität
Soldmannstr. 23
17489 Greifswald
Phone: +49 3834 864450
Fax: +49 3834 864451
E-Mail: fscholx@rz.uni-greifswald.de

Private address:
Lange Str. 87
17489 Greifswald

Domains of application:
Electrochemical analysis; solid state electrochemistry and electroanalysis; trace analysis, environmental analysis

Applied methods:
Electrochemical techniques; x-ray diffraction; STM-AFM; microscopic diffuse reflection spectroscopy; HPLC

Membership:
GDCh: Chairman of the division "Elektrochemische Analysenmethoden"; BPCh; International Society of Electrochemistry; Corresponding member of the International Academy of Computer Sciences and Systems

---

SCHUHMANN, Wolfgang, Prof. Dr.
Lehrstuhl für Analytische Chemie
Ruhr-Universität Bochum
Universitätsstr. 150
44780 Bochum
Phone: +49 234 7006200
Fax: +49 234 7094683

E-Mail:
woschu@anaserv.anachem.ruhr-uni-bochum.de

Domains of application:
Development of biosensors; microelectrochemistry; miniaturized sensors; Neurotransmitter and metabolite determination; bioprocess control; FIA

Applied methods:
Electroanalytical methods; cyclic voltammetry; scanning electrochemical microscopy; atomic force microscopy; amperometry with biosensors

Membership:
GDCh; MESEAP

SCHWEDT, Georg, Dipl.-Chem., Dr. rer. nat.
Institute of Inorganic and Analytical Chemistry
Technical University Clausthal
Paul-Ernst-Str. 4
38678 Clausthal-Zellerfeld
Phone: +49 5323 722209
Fax: +49 5323 722995
E-Mail: georg.schwed@tu-clausthal.de

Domains of application:
Environmental analysis; element speciation; sample preparation; food analysis

Applied methods:
Voltammetry; AAS, ion-chromatography; biotests; HPLC; electrophoresis; FIA

Membership:
RSC

SHELDRICK, William S., Dr.
Faculty of Chemistry
Ruhr-Universität Bochum
44780 Bochum
Phone: +49 234 7004192
Fax: +49 234 7094420
E-Mail: shel@anachem.ruhr-uni-bochum.de

Domains of application:
Metal-species analysis; bioorganometallic chemistry; structural analysis; crystal engineering

Applied methods:
XRD; NMR; MS; IR; UV/VIS; CD; HPLC; GC-MS; potentiometry

SIESLER, Heinz Wilhelm, Prof. Dr.
Department of Physical Chemistry
University of Essen
Schützenbahn 70
45117 Essen
Phone: +49 201 1832883
Fax: +49 201 1833967
E-Mail: hw.siesler@uni-essen.de

Domains of application:
Vibrational spectroscopy, mechanical measurements and thermal analysis of polymers; near-infrared light-fiber spectroscopy for chemical quality and process control; time-resolved step-scan FTIR-spectroscopy of liquid crystals in an external electric field

Applied methods:
FTIR and FT-Raman spectroscopy; thermal analysis

Membership:
Macromolecular Chemistry and Physics, Huthigh and Wepf Verlag; Journal of Near-Infrared Spectroscopy, NIR Publications

Awards:
1994 Eastern Analytical Sumposium Award in Near-Infrared Spectroscopy, USA

SPIEKERMANN, Manfred, Prof. Dr.
Fachhochschule Lübeck
Stephensonstr. 3
23562 Lübeck
Phone: +49 451 5005241
Fax: +49 451 5005236
E-Mail: spiekermann@fh-lubeck.de

Private address:
Schwalbenweg, 5
23628 Klempau

Domains of application:
Environmental analysis, food analysis, pharmaceuticals

Applied methods:
GC-FID; ECD; MS; WLD; HPLC-UV; fluorescence, RI; ion chromatography, FT-IR; AAS

Membership:
GDCh

SPITELLER, Michael, Prof. Dr.
Ecological Chemistry
University of Kassel
PO Box 1252
37202 Witzenhausen
Phone: +49 5542 981502
Fax: +49 5542 981504
E-Mail: spitell@wiz.uni-kassel.de

Domains of application:
Investigations according to OECD, BBA and EPA: degradation of chemicals in soil under an/aerobic conditions, photolysis of chemicals in water and on soil surfaces, hydrolysis of chemicals in aqueous buffer solutions

Applied methods:
Trace analysis of organic pollutants in water by HPLC-MS/MS and GC-MS; detection of C14-labelled compounds with LSC and image analyzer; laboratory lysimeters to study the leaching behaviour of organic pollutants; good laboratory practice (GLP)

Membership:
MESAEP

STEIN, Kathrin, Dipl. Chem., Dr. rer.nat.
Institute of Inorganic and Analytical Chemistry
Technical University Clausthal
Paul-Ernst-Str. 4
38678 Clausthal-Zellerfeld
Phone: +49 5323 722606
Fax: +49 5323 722895
E-Mail: kathrin.stein@tu-clausthal.de

Domains of application:
Enzymatic analysis; biochemical methods in food and environmental analysis; sample preparation

Applied methods:
Chromatography, CE; FIA; photometry

STÜBEN, Doris, Prof. Dr. rer.nat.
University of Karlsruhe
Kaiserstrasse 12
76128 Karlsruhe
Phone: +49 721 6083322
Fax: +49 721 6077247
E-Mail: doris.stueben@bip-geo.uni-karlsruhe.de

Private address:
Jasminweg 15
76228 Karlsruhe

Domains of application:
Geochemistry of rock, soil and water; stable isotopes; geochemical prospection; special trace elements with environmental relevance: Cr, Co, Cu, REE, Zn, As, Sc, Mo, Cd, Tl, Pb. Bi; platin-group-elements; uraban geochemistry

Applied methods:
Analytical chemistry; lake restauration techniques water treatment

Membership:
Professorship for geochemistry, environmental geochemistry, marine geochemistry, ore geochemistry/geology; member of the Earth Science Panel of TMR; Editorial Board of Marine Geology

Awards:
Humboldt fellowship: 1991/92 at the Oceanographic Institute, Wellington, New Zealand

---

SZARGAN, Rüdiger, Prof. Dr.
Wilhelm-Ostwald-Institute of Physical and Theoretical Chemistry
Linnestr. 2
04103 Leipzig
Phone: +49 341 9736451
Fax: +49 341 9736399
E-Mail: szargan@rz.uni-leipzig.de

Domains of application:
Surface science and surface physical chemistry: electronic and geometric structure, chemical bonding, chemical reactivity and transport processes; analysis of composition, structure, bonding and morphology on micrometer and nanometer scale, nanotechnology

Applied methods:
Photoelectron spectroscopy (XPS, ESCA, UPS) using synchrotron radiation, Auger electron spectroscopy (AES), scanning tunneling microscopy (STM), scanning electron microscopy (SEM), scanning electrochemical microscopy (SECM), electron diffraction methods

---

UNGER, Klaus, Prof. Dr.
Institut für Analytische Chemie und Anorganische Chemie
Johannes Gutenberg-Universität Mainz
Düsbergweg 10-14
55099 Mainz
Phone: +49 6131 395745
Fax: +49 6131 392710
E-Mail: unger@ak-unger.chemie.uni-mainz.de

Domains of application:
Synthesis and development of packings and stationary phases for analytical and preparative HPLC, electro-chromatography, high throughput screening, proteomics by RD-HPLC

Applied methods:
HPLC (analytical, preparative), CEC, RD-HPLC for protein mapping, miniaturized HPLC and CEC systems

Membership:
GDCh, member of the Working Group Chromatography, Federation of European Zeolite Organization (FEZA), Treasurer; AA-Zeolithe Decuema, Editorial Board member of J. Chromatogr., Chromatographia

Awards:
Pregl medal 1991, A.T.P. Martin Award Chromatographic Society, London, 1993; ACS Award in Chromatography 1995; Humboldt Professorship, Lund, Sweden, 1993

VON WACHTENDONK, Hans-Jürgen, Dr. rer.nat.
Chemische Laboratorien
Thyssen Stahl AG
Kaiser-Wilhelm-Strasse 100
47166 Duisburg
Phone: +49 203 5244986
Fax: +49 203 5228723

Domains of application:
Production control; certification of reference material; analysis of metals (mainly iron and steel) and inorganic materials; wateranalysis; gasanalysis; analysis of solid fuel (coal; coke, etc.)

Applied methods:
Spectrometry (XRF, XRD, AED, ICP, ICP-MS, GDOS, AAS); combustion method for analysis of non-metals in metals; gaschromatography with different detectors; HPLC; elemental analysis (coal, coke, oil)

Membership:
GDCh; Verein Deutscher Eisenhüttenleute

WENCLAWIAK, Bernd W., Prof. Dr. rer.nat.
Analytische Chemie I
University of Siegen
Adolf-Reichwein-Str. 9
57068 Siegen
Phone: +49 271 7404573
Fax: +49 271 7402414
E-Mail: wenclawiak@chemie.uni-siegen.d400.de

Private address:
Krottorfer Str. 77
57258 Freudenberg

Domains of application:
Inorganic analytical chemistry: metal speciation (Th, As, Hg); organic analytical chemistry: PAHs, pyrethroides, diesel gases
Applied methods:
HPLC; GC-MS; SFE; SFC; electrochemistry; AAS; ICP; IC; CE

Membership:
GDCh; AK Chromatographie; AMSEL; ASMS; The Engineering Society for Advancing Mobility and Air and Space); ACS; VDEH (Verein Deutscher Eisen-Hüttenleute)

WITTHAKER, Jürgen, Dr. rer.nat.
Institute of Occupational, Social and Environmental Medicine
Friedrich-Schiller-University Jena
Gustav-Freytag-Str. 1
99096 Erfurt
Phone: +49 361 7412482
Fax: +49 361 7412188
E-Mail: witthaker@ard.ef.uni-jena.de

Private address:
Steiger Str. 23d
99096 Erfurt

Domains of application:
Active and passive sampling in: outdoor air (VOC, NO_2, SO_2, et al.), indoor air (VOC, aldehydes, NO_2, et al.); bioacids in building materials and interiors

Applied methods:
Gas chromatography (FID, ECD, MS); HPLC; spectrophotometry

Membership:
GDCh; International Society of Indoor Air Quality and Climate

Awards:
Research Award of the Medical University Erfurt "Hygienic problems of indoor climatology", 1987

WOLFBEIS, Otto, Prof. Dr.
Institute of Analytical Chemistry, Chemo- and Biosensors
University of Regensburg
93040 Regensburg
Phone: +49 941 9434066
Fax: +49 941 9434064
E-Mail: otto.wolfbeis@chemie.uni-regensburg.de

Domains of application:
Chemical sensors; biosensors; fiber optic sensors for clinical applications; enzyme-based biosensors; immunosensors

Applied methods:
Fluorescence intensity; fluorescence lifetime; fluorescence polarization; energy transfer; fiber optic sensing; microsensors; immobilized indicators

Membership:
Society of Photo-Instrumentation Engineers; Society of Fluorescence

Awards:
1982: Sandoz-Prize; 1987: Feigl Prize for Microanalysis; 1989: Merck Prize; 1996: Emich Prize

---

ZSCHUNKE, Adolf, Prof. Dr.
Dept. Analytical Chemistry; Reference Materials
BAM
Rudower Chaussee 5
12489 Berlin
Phone: +49 30 63925800
Fax: +49 30 63925972
E-Mail: adolf.zschunke@bam.de

Private address:
Peitzer Weg 10
12527 Berlin

Domains of application:
Certification of reference materials; validation of analytical methods in materials testing, soil-analysis, water analysis; primary standards and traceability; proficiency testing; standardization

Applied methods:
Spectroscopical methods: AAS, AES, ICP-MS, x-ray fluorescence, FT-IR, Raman, Mössbauer; NMR; MS; time resolved fluorescence; chromatography; GC, CE, HPLC; x-ray diffraction: powder, single crystal; imaging methods; CT, micro-CT, NMR; nuclear methods

Membership:
EURACHEM, full committee member

Awards:
Research-prize of the Martin Luther University of Halle-Wittenberg; Friedrich Wöhler prize (Chem. Soc. of the former GDR)

ZWANZIGER, Heinz W., Prof. Dr.
Fachhochschule Merseburg
University of Applied Sciences
Geusär Strasse
06217 Merseburg
Phone: +49 3461 462027
Fax: +49 3461 462192
E-Mail: zwanziger@cni.fh-merseburg.de

Domains of application:
Education (national and international); environment; industry

Applied methods:
Automatic titration; electroanalytical (diverse); FAAS (FIAS); VIS-spectroscopy (DA); IR- and NIR-spectroscopy (several systems); chromatography (TLC; HPLC; IC; GC); chemometrics

Membership:
GDCh

Greece

ANDREDAKIS, George, Bc. Sci.
Department of Chemistry
University of Crete
71409 Iraklion
Phone: +30 81 393380
Fax: +30 81 210951

Greece

Private address:
Ippokratous, 37
71305 Iraklion

Domains of application:
Electroanalytical chemistry

Applied methods:
Sensors, ISE, optodes

ARCHONTAKI, Helen, Dr.
Department of Chemistry
University of Athens
Panepistimiopolis
15771 Athens
Phone: +30 1 7231608
Fax: +30 1 7231608

Private address:
Zupetis, 12
16121 Athens

Domains of application:
Laser applications; pharmaceutical and biomedical analysis; derivative UV/VIS spectrophotometry; HPLC

Applied methods:
Laser-induced breakdown spectroscopy; chemical stability of drugs (studied by derivative UV/VIS spectrophotometry and HPLC)

Membership:
Society for Applied Spectroscopy; New York Academy of Sciences

---

CALOKERINOS, Antony C., Dr.
Department of Chemistry
Laboratory of Analytical Chemistry
Panepistimiopolis
15771 Athens
Phone: +30 301 7284316
Fax: +30 301 7231608

Private address:
Thoukididou, 3-5
17455 Alimos, Athens

Domains of application:
Analytical chemistry; flame chemistry; ozone reactions

Applied methods:
Chemiluminescence

CHANIOTAKIS, Nikilas A., Dr.
Department of Chemistry
University of Crete
71409 Iraklion
Phone: +30 81 393137
Fax: +30 81 210951

Private address:
St. Nilolaidi, 31
71305 Iraklion

Domains of application:
Electroanalytical chemistry

Applied methods:
Sensors, biosensors, HPLC

Membership:
ACS

EFSTATHIOU, Constantinos, Dr.
Department of Chemistry
University of Athens
Panepistimiopolis
15771 Athens
Phone: +30 301 7284312
Fax: +30 301 7231608

Private address:
56 Kaftatzoglou Str.
11144 Athens

Domains of application:
Electroanalytical chemistry; chemometrics; reaction rate methods

Applied methods:
Applications of microcomputers in analytical instrumentation

EVMIRIDIS, Nicholaos, Dr.
Department of Chemistry
University of Ioannina
Panepistimioupolis/Dourouti
45110 Ioannina
Phone: +30 651 98404
Fax: +30 651 44112

Private address:
AG. Sikelianou, 2
45000 Ioannina

Domains of application:
Analytical chemistry; kinetics of chemical reactions; inorganic chemistry; surface methods of analysis; environmental chemistry; structure chemistry; electrochemistry; photochemistry; biochemistry; spectroscopy; membrane technology

Applied methods:
Synthesis and characterization of porous materials to be used as catalysts or ion-exchangers/adsorbents; development of analytical methods for a quick easy, sensitive and selective determination of chemical compounds; design of miniature sensors for rapid

Membership:
Chemometrics Society

GAVALAS, Vasilis, Bc. Sci.
Department of Chemistry
University of Crete
71409 Iraklion
Phone: +30 81 393135
Fax: +30 81 210951

Private address:
Delimarku, 11
71202 Iraklion

Domains of application:
Electroanalytical chemistry

Applied methods:
Biosensors

IOANNOU, Pinelopi, Dr.
Laboratory of Analytical Chemistry
University of Athens
Panepistimiopolis
15771 Athens
Phone: +30 1 7231608
Fax: +30 1 7231608

Private address:
Trifilios, 42
15772 Zogrofou

Applied methods:
Fluorimetry; pharmaceutical analysis; clinical chemistry; FIA; immunoassays

---

KANIAS, George, Dr.
Institute of Physical Chemistry
National Centre for Scientific Research "Demokritos"
Aghia Paraskevi Attikis
15310 Athens
Phone: +30 1 6513112
Fax: +30 1 6511766

Private address:
Xenofontos Street, 20
17673 Athens
Domains of application:
Analytical methods for the determination of trace elements in different media

Applied methods:
Neutron activation analysis and its application in the nutrition, environment, medicine, pharmaceutical sciences

---

KARAYANNIS, Miltiades, Dr.
Department of Chemistry
Laboratory of Analytical Chemistry
University Campus

45110 Ioannina
Phone: +30 651 98406
Fax: +30 651 98407

Private address:
Platia Omirou 3
45332 Ioannina

Domains of application:
Flow injection analysis; kinetics analytical chemistry; electroanalysis; gas- and liquid-chromatography; environmental analysis; chemometrics

Applied methods:
Biosensing; FIA-analysis

Membership:
ACS; MESAEP; FECS; EURACHEM

KONIDARI, Constantina, Dr.
Department of Chemistry
University of Ioannina
University Campus
45110 Ioannina
Phone: +30 651 98408
Fax: +30 651 98407

Private address:
Pedini Ioanninou
45000 Ioannina

Domains of application:
Chemical kinetics; environmental chemistry

Applied methods:
Chemical kinetics

Membership:
MESAEP

KOUPPARIS, Michael, Dr. Sci.
Laboratory of Analytical Chemistry
University of Athens
Panepistimiopolis
15771 Athens
Phone: +30 1 7284559
Fax: +30 1 7231608

Private address:
Peloponnisou, 18
15235 Viilisia, Attika

Applied methods:
Pharmaceutical analysis; QC of drugs; kinetic method of analysis; potentiometric sensors (ion-selective electrodes); micellar catalysis; automated methods of analysis; FIA; HPLC; immunoassays

Membership:
ACS; AOAC; European Pharmacopeia

LAFIS, Spyros, Dr. Sci.
Laboratory of Analytical Chemistry
University of Ioannina
45110 Ioannina
Phone: +30 651 98412
Fax: +30 651 98407
E-Mail: slafis@cc.uoi.gr

Domains of application:
Analytical chemistry; sensor development; kinetic studies

Applied methods:
Electrochemistry; optical methods; stopped-flow techniques

LIODAKIS, Stelios, Dr.
National Technical University of Athens
Iroon Polytechniou, 19
15773 Athens
Phone: +30 1 7723116
Fax: +30 1 7723188

Private address:
Zitsis, 9
15123 Marussi Polydrosso

Domains of application:
Instrumental methods of analysis;

chromatography; thermal analysis; spectrometry

Applied methods:
GC; HPLC; analysis of essential oils; analysis of pyrolysis products of forest fuels

---

MOSCHOU, Elizabeth,
University of Crete
71409 Iraklion
Phone: +30 81 333135
Fax: +30 81 210951

Private address:
Solonos, 265, Fortetsa
71305 Iraklion

Domains of application:
Electroanalytical chemistry

Applied methods:
Sensors; ISEs; ISFETs

NANOS, Christos, Dr.
Department of Chemistry
University of Ioannina
University Campus
45110 Ioannina
Phone: +30 651 98405
Fax: +30 651 98407

Private address:
Pedini Ioanninou
45000 Ioannina

Domains of application:
Electrochemistry; environmental chemistry; biosensors
Applied methods:
Electrochemistry

Membership:
MESAEP

OCHSENKUEHN, Klaus-Michael, Dr. Chem.
Institute of Physical Chemistry
National Center for Scientific Research "Demokritos"
Aghia Parashevi
15310 Athens
Phone: +30 1 6513111
Fax: +30 1 6511766

Private address:
Trifylias 4b
15231 Athens

Domains of application:
Trace element analysis; speciation analysis; environmental analysis; geochemistry; archaeometry

Applied methods:
Neutron activation analysis; AAS

Membership:
GDCh; Association of Geeek Chemists; Greek Archaeometric Society

---

OCHSENKUEHN-PETROPULU, Maria, Assoc. Prof. Dr.
Laboratory of Analytical and Inorganic Chemistry
National Technical University of Athens
Heroon Polytechniou, 9
15773 Athens
Phone: +30 1 7723094
Fax: +30 1 7723188

Private address:
Trifilias, 4b
15231 Athens

Domains of application:
Speciation analysis; trace and ultratrace element analysis; preparation and characterization of high temperature superconductors, powders and contings; environmental analysis; determination and recevory of lanthanides from industrial by products

Greece

Applied methods:
Development of analytical methods for speciation analysis; separation and determination of trace elements from environmental matrices

Membership:
Deutsche Gesellschaft für Chemischen Apparatebau; IUPAC; Chemical Society for Thermal Analysis; International Confid. of Thermal Analysis and Calorimetry

---

PAPADOYANNIS, Ioannis, Dr.
Laboratory of Analytical Chemistry
University of Thessaloniki
54008 Thessaloniki
Phone: +30 31 997793
Fax: +30 31 997719

Private address:
Philippou, 27
54631 Thessaloniki

Domains of application:
HPLC, solid phase extraction; atomic absorption spectroscopy; ion chromatography

Applied methods:
The analysis of components with pharmaceutical and biological interest; the determination of inorganic components in different samples

Membership:
Forensic Sciences Society; Chromatographic Society; Union of Greek Chemistry

---

PAPASTATHOPOULOS, Demetrius, Dr. Sci.
Department of Chemistry
Laboratory of Analytical Chemistry
Panepistimiopolis
15771 Athens
Phone: +30 301 7231608
Fax: +30 301 7231608

Private address:
Artemisiou, 15
15231 Halandri

Domains of application:
Ion selective electrodes; enzyme biosensors; automated methods of bioanalysis; enzyme isolation and purification; immobilization of enzymes

Applied methods:
Enzyme biosensors using immobilization techniques; enzymatic analysis

---

PAPPA, Athina, Lecturer
National Technical University of Athens
Iroon Polytechniou, 19
15773 Athens
Phone: +30 1 7723193
Fax: +30 1 7723188

Private address:
Agerthonos, 16
15343 Athens

Domains of application:
Instrumental methods of analysis; chromatography; HPLC; gas chromatography; thermal analysis; TG; DSC; hyphenated techniques

Applied methods:
Pyrolysis of ping-meedles; study of the effect of chemicals of in

---

PARISSAKIS, George, Prof. Dr.
Laboratory of Analytical and Inorganic Chemistry
National Technical University of Athens
Heroon Polytechniou
15773 Athens
Phone: +30 1 7723100
Fax: +30 1 7723188

Private address:
Niridon, 2
17561 Phaliron

Domains of application:
Gas chromatography; HPLC; mass spectrometry; recovery of elements from industrial water; element chemistry and technology; powder metallurgy; thermal analysis

Applied methods:
Gas chromatography; HPLC, mass spectrometric; element chemistry and technology; recovery of elements from screep; fire retardants; thermal analysis

Membership:
Gesellschaft Thermalyzer Polytechniker; International Pollution Risk Advisers; European Element Association; Club of Athens; IUPAC

PILIDIS, George, Dr.
Department of Chemistry
University of Ioannina
University Campus
45110 Ioannina
Phone: +30 651 98391
Fax: +30 651 98381

Private address:
Chimaras, 16-20
45445 Ioannina

Domains of application:
Determination of organic pollutants in soil, water, air, textiles, olive oils; synthesis of organic compounds

Applied methods:
Pesticides, phenols, PAHs, VOCs

Membership:
MESAEP; GDCh; ACS

PIPERAKI, Efrosini, Dr. Sci.
Department of Chemistry
University of Athens
Panepistimiopolis
15771 Athens
Phone: +30 1 7284556
Fax: +30 1 7231608

Private address:
Thiseos, 3
16345 Athens

Domains of application:
Spectroscopy; trace metal analysis

Applied methods:
AAS; chemical modification for the determination of trace elements by electrothermal atomization; AAS in various samples

---

SAMANIDOU, Victoria, Dr. Sci.
Laboratory of Analytical Chemistry
University of Thessaloniki
54006 Thessaloniki
Phone: +30 31 997698
Fax: +30 31 997719

Private address:
Anixeos, 39
55236 Thessaloniki-Panorama

Domains of application:
HPLC, IC, GC SPE

Applied methods:
Application of above mentioned techniques in the analysis of organic and inorganic compounds in samples of pharmaceutical biological and environmental interest

Membership:
Union of Greek Chemistry; Chromatographic Society

SISKOS, Panayotis A., Dr.
Department of Chemistry
University of Athens

Panepistimiopolis
15771 Athens
Phone: +30 301 7284311
Fax: +30 301 7231608

Domains of application:
Chromatographic methods; kinetic methods of analysis; environmental analysis; clinical analysis

Applied methods:
Analytical chemistry; environmental analysis; clinical analysis

Membership:
ACS; The International Society of the Build Environment

STATHEROPOULOS, Miltiades, Assist. Prof.
Department of Chemical Engineering
National Technical University of Athens
Iroon Polytechniou, 19
15773 Athens
Phone: +30 1 7723199
Fax: -30 1 7723188

Domains of application:
Chemometrics; plastics analysis; software development (expert systems); forest fire chemicals; mobile analytical instruments

Applied methods:
Development of analytical methods

Membership:
ACS

THEMELIS, Demetrius G., Dr. Sci.
Laboratory of Analytical Chemistry
Aristotle University of Thessaloniki
54006 Thessaloniki
Phone: +30 31 997804
Fax: +30 31 997719

Private address:
Filikid Eterias, 15
54621 Thessaloniki

Domains of application:
Kinetic methods of analysis; water process control; automation
Applied methods:
Flow injection analysis

TSAGATAKIS, Ioannis, Bc. Sci.
Department of Chemistry
University of Crete
71409 Iraklion
Phone: +30 81 393380
Fax: +30 81 210951

Private address:
Knossou, 143
71306 Iraklion

Domains of application:
Electroanalytical chemistry; HPLC

Applied methods:
Sensors, ISEs; biosensors

TZAMTZIS-PILALIS, Nikolaos, Lecturer
National Technical University of Athens
Iroon Polytechniou, 19
15773 Athens
Phone: +30 1 7723194
Fax: +30 1 7723188
Private address:
L. Marathonos, 171; 15344 Athens

Domains of application:
Analytical chemistry; applications; instrumentations

Applied methods:
Gas chromatography; mass spectrometry; pyrolytic methods; chemometrics on GC

VOULGAROPOULOS, Anastasios, Dr.
Aristotle University of Thessaloniki
University Campus
54006 Thessaloniki
Phone: +30 31 997376
Fax: +30 31 997719

Private address:
Ammolianis Str. 9
54638 Thessaloniki
Domains of application:
Electroanalytical chemistry; chemical metrology

Applied methods:
Trace analysis in environmental and biological samples; metrology in chemistry

Membership:
AAAS

ZACHARIADIS, George, Dr.
Laboratory of Analytical Chemistry
Aristotle University of Thessaloniki
54006 Thessaloniki
Phone: +30 31 997707
Fax: +30 31 997719

Private address:
Peridlious, 11
55006 Thessaloniki

Domains of application:
Analytical chemistry

Applied methods:
Chemical analysis: spectroscopic methods; heavy metals in the environment, chemometrics

Membership:
Greek Chemical Association

ZOTOU, Anastasia, Dr.
Laboratory of Analytical Chemistry
Aristotle University of Thessaloniki
54006 Thessaloniki
Phone: +30 31 997746
Fax: +30 31 997719

Private address:
Koritsas Str. 3, Agies Parlos
55438 Thessanoliki

Domains of application:
High pressure liquid chromatography and capillary electrophoresis separation and determination of various compounds (pharmaceuticals, biological molecules, toxins, ion, etc)

Applied methods:
High pressure liquid chromatography; high performance capillary electrophoresis

Membership:
Chromatographic Society of Britain

Hungary

BACSA, Gyorgy, Dr.
Chemical Works of Gedeon Richter Ltd.
Gyomroi ut 19-21
1475 Budapest
Phone: +36 1 4315592
Fax: +36 1 2624066
E-Mail: h6853.bac@ella.hu

Private address:
Ond Vezer u. 17a
1144 Budapest

Domains of application:
Total quality management for: research, development, manufacture and sales of pharmaceutical; active ingredients and final products

Applied methods:
Quality control; quality assurance; audits; quality improvement

Membership:
Hungarian Pharma Section of the European Organization of Quality; International Who's Who of Intellectuals Cambirdge, England; American Biographical Institute

Awards:
Diploma from the Hungarian Minister of Foreign Trade, 1986; Diploma from the Hungarian Minister of Heavy Industry, 1979; Worls Decoration of Excellence ABI, 1989; Excellent Innovator, Gold, 1985; Spergely Bela Commemorative Medal, 1990

BALOCH, Sandor, Dr. Sci.
Reanal Factory of Fine Chemistry Co. Ltd.
Telepes u. 53
1147 Budapest
Phone: +36 1 2513999
Fax: +36 1 2521339

Private address:
Ratz Laszlo u. 58
1116 Budapest

Domains of application:
Quality control; quality assurance

Applied methods:
Standard analytical methods; inorganic and organic; instrumentation: AAS, GC, HPLC, UV/VIS, IR, spectrometry NMR

BARCZA, Lajos, Dr. Sci.
Department of Inorganic and Analytical Chemistry
L. Eotvos University
PO Box 32
1518 Budapest
Phone: +36 1 2090555
Fax: +36 1 2090602
E-Mail: barcza@ludens.elte.hu

Domains of application:
Teaching in analytical chemistry, inorganic chemistry and supramolecular chemistry; research in above mentioned fields

Applied methods:
Spectrophotometry; electroanalytical methods; HPCE

Awards:
Schulek Prize

BEZUR, Laszlo, Dr.
Institute for General and Analytical Chemistry
Technical University of Budapest
Set. Gellert Ter. 4
1521 Budapest
Phone: +36 1 4631178
Fax: +36 1 4633408
E-Mail: bez@ch.bme.hu

Private address:
Bartok B. u. 3a, 1225 Budapest

Domains of application:
Analytical chemistry; research and development in electroanalytical chemistry; chromatography; thermal analysis; AAS; IR spectroscopy; NMR

Applied methods:
AAS, flame-AAS, ETA-AAS, Hg-hydrid-AAS, ICP-OES, thermo-AAS

Membership:
IUPAC

BIACS, Peter Akos, Dr. techn., Chem. Eng.
Central Food Research Institute
Herman Otto u. 15
1022 Budapest
Phone: +36 1 1558991
Fax: +36 1 1558991
E-Mail: h9740bia@ella.hu

Private address:
Bartok Bela ut 15/C
1114 Budapest

Domains of application:
Food analysis (chemical, physical, biological); biotechnology (fermentation and enzyme application and analysis); food packaging (DIA, permeability, pheology); microbiological analysis/assay

Applied methods:
Chromatography (HPLC, GC-MS, TLC) of lipids and lipoids (carotenoids, tocopherols, waxes); spectroscopy (UV, IR, NMR, AAS); sampling and separation

Membership:
International Union of Food Science and Technology; International Life Science Institute European Branch; Member of the Board of Directors

Awards:
FAO Certificate of Merit for Codex Alimentarius Activity

BURGER, Kalman, Dr. Sci.
Department of Inorganic and Analytical Chemistry
A. Jozsef University
Dom ter 7
6720 Szeged
Phone: +36 62 454054
Fax: +36 62 312505

Private address:
Somogyi u. 6
6720 Szeged

Domains of application:
Solution chemistry; coordination chemistry; instrumental analysis

Applied methods:
Coordination equilibria; structural study of coordination compounds; Mössbauer spectroscopy

Membership:
Accademiz Peloritana dei Pericolanti

BUVARI-BARCZA, Agnes, C. Sci.
Department of Inorganic and Analytical Chemistry
L. Eotros University
PO Box 32
1518 Budapest
Phone: +36 1 2090555
Fax: +36 1 2090602
E-Mail: buvari@ludens.elte.hu

Domains of application:
Teaching in analytical (and inorganic) chemistry; research in analytical chemistry, hydrogen bonding and complex equilibria, inclusion phenomena

Applied methods:
Spectrophotometry; fluorescence; capillary electrophoresis; electrochemical methods

---

DINYA, Zoltan, Dr. Sci.
Department of Chemistry
University of Debrecen
University Square, 1
4010 Debrecen
Phone: +36 52 316666
Fax: +36 52 310936
E-Mail: dinya_z@tigris.klte.hu

Private address:
Komlossy str. 28
4028 Debrecen

Domains of application:
Trace organic analysis; environmental organic analysis; structure determination of natural products (antibiotics, flavonoids, carbohydrates); identification of drug metabolites

Applied methods:
GC; LC; MS; GC-LC-MS; UV-VIS, FT-IR, Raman)

Membership:
IUPAC

Awards:
Master Teacher "Szechenyi Professzori Eratendij"

FALKAY, George, Dr. Sci.
Department of Pharmacodynamics
Albert Szent-Gyorgy Medical University
Eotvos u. 6
6720 Szeged
Phone: +36 62 455568
Fax: +36 62 455567
E-Mail: falkay@pharma.szote.u-szeged.hu

Private address:
Kallay A. U. 9/A
6726 Szeged

Domains of application:
Therapeutic drug monitoring; pharmacokinetic studies; characterization of adrenergic; steroide receptors by radioligand binding techniques

Applied methods:
Radioimmunoassay; receptorassay; HPLC analysis of drugs in biological matrix

Membership:
International Federation of Clinical Chemistry

FEHER, Zsofia, Dr. Sci.
Institute for General and Analytical Chemistry
Technical University of Budapest
Szt. Gellert Ter. 4
1111 Budapest
Phone: +36 1 4631592
Fax: +36 1 4633408
E-Mail: feher@ch.bme.hu

Private address:
Hegedis Gy, 49-51
1013 Budapest

Domains of application:
Analytical chemistry; different kinds of instrumental analytical techniques: AAS, electroanalysis, separation techniques, MS, IR spectrometry, NMR technique, thermal analysis, analysis of molecular structures

Applied methods:
Electroanalytical techniques; sensor-research; flow through analysis

Awards:
Award of the Hungarian Academy of Sciences

FEKETE, Jeno, Dr.
Institute of General and Analytical Chemistry
Technical University of Budapest
Szt. Gellert Ter. 4
1111 Budapest
Phone: +36 1 4631596
Fax: +36 1 4633408
E-Mail: fekete@ch.bme.hu

Private address:
Kovidinka u. 20
1028 Budapest

Domains of application:
Environmental analysis; drug analysis; residue and metabolite determination; chiral separation

Applied methods:
HPLC; GC

FERENCZ, Antal, Dr.
Institute of Clinical and

Experimental Laboratory Medicine
Imre Haynal University of Health Sciences
Szabolcs u. 35
1135 Budapest
Phone: +36 1 2702211
Fax: +36 1 1208246

Private address:
Silvanus Setany 31
1031 Budapest

Domains of application:
Molecular biology; clinical microbiology; diabetology; laboratory medicine; analytical clinical chemistry

Applied methods:
Analytes in clinical chemistry; glycated proteins; advanced glycation endproducts; metabolites of carbohydrate and lipid metabolism; other analytes: urea, haemoglobin

Membership:
IUPAC; Societe Francaise de Biologie Chimique; Institut für Standardisierung und Dokumentation in Medizinischen Laboratorium E.V.; Deutsche Gesellschaft für Qualitätssicherung in GER Laboratoriumsmedizin E.V.

GOROG, Sandor, Dr. Sci.
Chemical Works of Gedeon Richter Ltd.
Gyomroi u. 21
1475 Budapest
Phone: +36 1 4314620
Fax: +36 1 2605604

Private address:
Valyog u. 2-4
1032 Budapest

Domains of application:
Pharmaceutical analysis; analytical aspects of pharmaceutical research and development; method development for the assay and impurity profiling of bulk drugs and drog formulations; stability assays; analytical aspects of metabolic studies

Applied methods:
Classical methods: titrations, spectrophotometry; chromatography: TLC, GC, HPLC; capillary electrophoresis; spectroscopic methods: UV, IR, Raman, MS, NMR; hyphenated methods: GC-MS; HPLC-UV; HPLC-MS

Membership:
FECS; International Society for Pharmaceutical and Biomedical Analysis; J. Pharm. Biomed Anal. Trends in Anal. Chem.

Awards:
Szechenyi-Prize, Than, Schulek, Szebelledy Medals

GOTH, Laszlo, Cand. of biological science
Department of Clinical Chemistry
Medical University of Debrecen
Nagyerdei Krt. 98
4012 Debrecen
Phone: +36 52 431956
Fax: +36 52 417631
E-Mail: goth@kki.dote.hu

Domains of application:
Clinical chemistry; diagnostic enzymology; molecular biology; acatalasemia

Applied methods:
Clinical chemistry analyses (spectrophotometric, potemtiometric, HPLC); molecular biology; PCR; SSCP; heteroduplex; sequencing

Membership:
New York Academy of Sciences;

American Association for Clinical Chemistry; International Society for Clinical Enzymology

Awards:
Pandy Kalman by Hungarian Society of Clinical Pathology; Fulbright Research Fellowship

---

HLAVAY, Jozsef, Dr. Sci.
Department of Earth and Environmental Science
University of Veszprem
Egyetem u. 10
8201 Veszprem
Phone: +36 88 423203
Fax: +36 88 421869
E-Mail: hlavay@anal.venus.vein.hu

Domains of application:
Analysis of environmental samples; investigation of aerosols; chemical sensors; separation methods; models for the speciation of chemical compunds

Applied methods:
GC; HPLC; CE; AAS; XRD; XRF

Membership:
IUPAC/V.2; EURACHEM

Awards:
Transdanubian Analytical Awards, 1986; Pro Aqua Award, 1993

HORVAI, George, Dr.
Division of Chemical Information Technology
Technical University of Budapest
Gellert Ter. 4
1111 Budapest
Phone: +36 1 4631480
Fax: +36 1 4633953
E-Mail: horvai@ch.bme.hu

Domains of application:
Electroanalysis; ion-selective electrodes; trace analysis; pharmacokinetic analysis; automated analysis

Applied methods:
Potentiometry; voltammetry; HPLC; analytical robots; flow injection analysis

Membership:
IUPAC; ISE

Awards:
Academy Prize of the Hungarian Academy of Science

HORVATH, Gyula, Ms. C., Dr. C. Sci.
Institute for Drug Research Ltd.
Berlini, 47-49
1045 Budapest
Phone: +36 1 1690011
Fax: +36 1 1693229

Private address:
V. Kigyo u. 4-6, 1052 Budapest

Domains of application:
Analysis of synthetic and natural organic compunds; drug candidates; drug substance impurities; metabolites

Applied methods:
Titrations; elemental analysis; purity controls by HPLC, TLC, GC; structural analysis (UV, IR, NMR, MS)

Membership:
International Mass Spectrometry Conference

HORVATH, Viola, Dr.
Institute of General and Analytical Chemistry
Technical University of Budapest
Gellert Ter. 4
1111 Budapest
Phone: +36 1 4634050

Fax: +36 1 4633408
E-Mail: v-horvath@ch.bme.hu

Domains of application:
Analytical Chemistry; electroanalytical chemistry; spectroscopic techniques; chromatography; structural chemistry

Applied methods:
HPLC; GC; MS; IR; AAS; potentiometry; voltammetry; thermal analysis

---

IDEI, Miklos, Dr.
Department of Medical Chemistry, Molecular Biology and Pathobiochemistry
Semmelweis University of Medicine
Puskin u. 9
1088 Budapest
Phone: +36 1 2662755
Fax: +36 1 2667480
E-Mail: miki@puskin.sote.hu

Private address:
Karpat u. 16
1088 Budapest

Domains of application:
Education: graduate and post graduate courses in the Medical University; research: in the field of biochemistry, pathobiochemistry, molecular biology, drug development, separation techniques

Applied methods:
High performance liquid chromatography; capillary electrophoresis; micellar electrokinetic chromatography; preparative chromatography

Awards:
Award of the Hungarian Biochemical Society; Award of the Semmelweis Medical University for junior lecturers; Award of the International Tswett Foundation

---

INCZEDY, Janos, Dr. techn., Dr. Sci.
Institute of Analytical Chemistry
University of Veszprem
Egyetem u. 10
8200 Veszprem
Phone: +36 88 422022
Fax: +36 88 421869
E-Mail:
inczedy@anal.venus.vein.hu

Private address:
Budafohi ut. 45
1111 Budapest

Domains of application:
Analytical Chemistry

Applied methods:
Ion exchange; process analytical methods

Membership:
FECS: Member of Advisory Board; IUPAC: Member of Commission V.1

Awards:
Hungarian State 1980; Than (Hung. Chem. Soc) 1977; Schniek Medal (Hung. Pharm. Soc.) 1987; Gold Medal of Univ. Veszprem 1988; Corresp. Member of H. Acad. Sci 1993; Prof. Emeritus (Veszprem Univ.) 1995

INZELT, Gyorgy, Dr. Sci.
Department of Physical Chemistry
Eotvos University
PO Box 32
1518 Budapest
Phone: +36 1 2090555
Fax: +36 1 2090602
E-Mail: inzeltgy@ludens.elte.hu

Private address:
Attila ut. 39

1013 Budapest

Domains of application:
Physical chemistry; electrochemistry and electroanalytical chemistry; polymer film electrodes; electrocatalysis; electrosorption; electrochemical oscillations; biosensors; sensors; electrochromic displays

Applied methods:
Cyclic voltammetry; potentiometry; electrochemical quartz crystal microbalance; impedance spectroscopy; radiotracer technique; spectroelectrochemistry (UV-VIS, ESR); amperometry

Membership:
ISE; IUPAC; FECS

Awards:
Award for Excellence in Teaching and Research, 1985; University Prize or Science, 1991

KALASZ, Huba, Dr. Sci. (Pharmacology)
Department of Pharmacology
Semmelweis University of Medicine
Negyvarad ter. 4
1445 Budapest
Phone: +36 1 2102930
Fax: +36 1 2104412
E-Mail: kalhub@net.sote.hu

Private address:
Gvadanyi u. 44-46
1144 Budapest

Domains of application:
Pharmacology, metabolic studies, pharmacokinetics; isolation and structural elucidation of natural products; discovery of new metabolites; studies on chiral discrimination during metabolism; receptor binding sites

Applied methods:
Thin-layer chromatography; HPLC; CE; displacement chromatography

Membership:
The New York Academy of Science; Chromatography Discussion Group of the Hungarian Chemical Society

Awards:
New England Chromatography Council, Award, 1985

KOVACS, Gabor L., Dr. Sci.
Central Laboratory
Markusovszhy Teaching Hospital
Markusovszky Str. 3
9701 Szombathely
Phone: +36 94 312172
Fax: +36 94 327873

Private address:
Fogaras u. 37
9700 Szombathely

Domains of application:
Clinical chemistry; endocrinology; immunology; laboratory financing

Applied methods:
Laboratory medicine; neuroendocrinology; management

Membership:
International Society of Neuroendocrinology; New York Academy of Sciences; Alpe Adria Laboratory Organization; Hungaria Society of Clinical Pathology

---

KOVACS-HADADY, Katalin, Cand. Sci.
Department of Inorganic and Analytical Chemistry
L. Kossuth University
PO Box 21
4010 Debrecen
Phone: +36 52 316666
Fax: +36 52 310936

E-Mail: hadady@tigris.klte.hy

Domains of application:
Education (instrumental analysis, lectures and laboratory practice, electrochemistry, planar chromatographic techniques, CE); research development of new methods: study of retention mechanism on impregnated thin-layers

Applied methods:
Overpressured layer chromatograhy; capillary electrophoresis; electroanalytical techniques; spectrophotometry

---

KREMMER, Tibor, Dr. Sci.
Department of Biochemistry
National Institute of Oncology
Rath Gy. Str. 5-7
1125 Budapest
Phone: +36 1 555682
Fax: +36 1 562402
E-Mail: kremmer@cncol.hu

Domains of application:
Biochemistry; bioanalysis; analytical (organic) chemistry; separation sciences; chromatography; clinical chemistry; cancer research

Applied methods:
Bioanalysis of natural compounds; chromatography; electrophoresis; ultracentrifugation; spectrophotometry; lyophilization

Membership:
Association of Cancer Research; Hungarian Committee

KRISTOF, Janos, Dr. Sci.
Department of Analytical Chemistry
University of Veszprem
Egyetem u. 10
8201 Veszprem
Phone: +36 88 421869
Fax: +36 88 421869
E-Mail:
kristof@anal.venus.vein.hu

Private address:
Vajda J.u. 10
8200 Veszprem

Domains of application:
Retention theory of ion chromatography; analytical application of surface enhanced Raman spectrometry; surface studies of thin films and catalyst systems; air pollution analysis by FT-IR spectrometry

Applied methods:
IC; HPLC; FT-IR; FI-Raman; SERS; TG-MS; TG-DTG-DIA; SIMS; DRIFT; high rresolution GT-IR with long path mos cells; ICP-OES; AAS

LAJOS, Papp, Dr. Sci.
Department of Inorganic and Analytical Chemistry
Lajos Kossuth University
Egyetem Ter. 1
4010 Debrecen
Phone: +36 52 316666
Fax: +36 52 310936
E-Mail: mathee@tigris.klte.hu

Private address:
Tessedik u. 183, 4032 Debrecen

Domains of application:
Trace element analysis of environmental, biological and food samples

Applied methods:
FAAS, GFAAS, HG-QF-AAS, ICP-AES, ETA-HCL-OES

LEMBERKOVICS, Eva, Dr.
Institute of Pharmacognosy
Semmelweis University of Medicine
Ulloi

1085 Budapest
Phone: +36 1 2660120
Fax: +36 1 1172979
E-Mail: lembi@drog.sote.hu

Private address:
Ulloi street, 66
1082 Budapest

Domains of application:
Research of natural plant substances phytochemical analysis, isolation, structure elucidation; developing of complex plant preparation for therapeutical use

Applied methods:
Spectroscopy (IR, NMR, UV); chromatography (SC, HPLC, TLC, HPTLC, OPLC, GC, GC-IR); extraction SFE; water steam distillation; solvent extraction

Membership:
International Federation of Essential Oils and Aroma Trades (ISFEAT)

LINDNER, Erno, Dr. Sci.
Department of Chemical Engineering
Technical University of Budapest
Szt. Gellert Ter. 4
1111 Budapest
Phone: +36 1 4631288
Fax: +36 1 4634032
E-Mail: lidner@ch.bme.hu

Domains of application:
Analytical chemistry

Applied methods:
Electroanalytical chemistry; chemical sensors; flow analytical techniques; microelectronically fabricated sensors

Membership:
IUPAC

Awards:
Award of the Hungarian Academy of Sciences, 1990; Award of the Best Scientific Paper of the year, 1989 by the Hungarian Journal of Chemistry; Golden Grade Award of Honnoured Inventor, 1986

LIPTAY, Gyorgy, Dr. Sci.
Institute for Inorganic Chemistry
Technical University of Budapest
Gellert Ter. 4
1521 Budapest
Phone: +36 1 4631479
Fax: +36 1 4633642

Private address:
Andrassy ut. 20, 1061 Budapest
Domains of application:
Chemical analysis

Applied methods:
Thermal analysis; simultaneous thermal analysis (TG, DTG, DTA); applications of thermal analysis

Membership:
European Symposium for Thermal Analysis and Calorimetry; IUPAC; European Society for Thermal Analysis, Calorimetry, Thermodynamic and Chemistry Reactivity

Awards:
Koroly THAN-Award of Hungarian Chemical Society

MARTON, Aurel, Dr. Sci.
Department of Analytical Chemistry
University of Veszprem
PO Box 158
8201 Veszprem
Phone: +36 88 422022
Fax: +36 88 421869
E-Mail: marton@anal.venus.vein.hu
Private address:
Vezpremvolgyi St. 17
8200 Veszprem

Domains of application:
Environmental analytical chemistry

Applied methods:
Electroanalytical; spectroscopic (NMR, IR); separation (GC, HPLC)

Membership:
IUPAC

MOLNAR-PERL, Ibolya, Dr. Sci.
Institute of Inorganic and Analytical Chemistry
L. Eotios University
PO Box 32
1518 Budapest
Phone: +36 1 2090608
Fax: +36 1 2090602
E-Mail: perlne@para.chem.eltp.hu

Private address:
Dobsinai Street 11
1124 Budapest

Domains of application:
Analytical method development; improvement in any field of analytical chemistry

Applied methods:
GC/MS; HPLC/UV/DAD, spectrometry UV/fluorescent

Membership:
New York Academy of Sciences

MOLNAR, Pal, Dr. Sci.
Central Food Research Institute
Herman Otto u. 15
1537 Budapest
Phone: +36 1 1565082
Fax: +36 1 2741005

Private address:
Hurosnalgy ut 157
1021 Budapest

Applied methods:
Sensory evaluation; ecological methods; pattern recognition of food analysis data

Membership:
European Organization for Quality; Codex Alimentarius Commission Methods of Analysis and Sampling; FECS

Awards:
For standardization and Quality, 1984

MORVAI-VITANYI, Magdolna, Dr.
Chinoin Pharmaceutical and Chemical Works Co. Ltd.
Istvantelki ut. 19
1325 Budapest
Phone: +36 1 1690900
Fax: +36 1 1690293

Private address:
Alkotmany u. 31
1054 Budapest

Domains of application:
Pharmaceutical chemistry; analytical chemistry; photochemistry; stability studies of drug substances and drug products; analytical method development

Applied methods:
GC; HPLC; coupled techniques (GC/MS, HPLC/MS); spectrophotometry

Awards:
Schulek E. Award, 1994

NAGY, Erzsebet, Dr. Sci.
Medical School of Debrecen
University of Debrecen
University Square, 1
4010 Debrecen
Phone: +36 52 316666
Fax: +36 52 310936

Private address:
G. Jeriko Str.
4032 Debrecen

Domains of application:
Trace organic analysis; environmental organic analysis; structure determination of natural products (antibiotics, feavowords, carbohydrates); identification of drug metabolites

Applied methods:
Liquid chromatography (LC), gas chromatography (GC), mass spectrometry (MS); combinated techniques (GC-LC-MS); optical spectrometric techniques (UV-VIS, IT-IR, Raman)

NAGY, Geza, Dr. Sci.
Institute of General and Analytical Chemistry
Technical University of Budapest
Szt. Gellert Ter. 4
1111 Budapest
Phone: +36 1 4631415
Fax: -36 1 4634052
E-Mail: g-nagy@ch.bme.hu

Domains of application:
Teaching analytical chemistry; research in the field of analytical chemistry; electroanalytical chemistry

Applied methods:
Potentiometry; voltammetry; scanning electrochemical microscopy; biosensors

Awards:
Academy Award, 1987

NOSZAL, Bela, Dr. Sci.
Institute of Pharmaceutical Chemistry
Semmelweis University of Medicine
Hogyes E. U. 9
1092 Budapest
Phone: +36 1 2170891
Fax: +36 1 2170891
E-Mail: nosbel@hogyes.sote.hu

Domains of application:
Analytical methods in pharmaceutical chemistry: speciation; microspeciation, separation techniques, spectroscopic techniques; synthesis and characterization of drug candidates; physico-chemical characterization (pKs; logP) of drugs and prodrugs

Applied methods:
Separation techniques: GC, HPCE, HPLC, TLC and hyphenated methods; spectroscopic techniques: UV-VIS, CD, ORD, IR, NMR

Awards:
A. Guzagh Award (Hungarian Academy of Sciences) in Physical Chemistry

---

NYIREDY, Szabolcs, Dipl. Pharm., Dr. Sci.
Research Institute for Medicinal Plants
Lupaszigeti ut 4
2011 Budakalasz
Phone: +36 26 340203
Fax: +36 26 340426

Private address:
Volgy u. 49
1021 Budapest

Domains of application:
Promoting research and technological development within the medicinal plant production and manufacturing of heral medicines facilitating the application of scientific advances

Applied methods:
Analytical and preparative separation techniques (extraction,

chromatography); genetic; biotechnology

Membership:
Hungarian Society for Separation Sciences; Foreign Affairs of teh Hungarian Pharmaceutical Society; Chromatographic Society; Int. Society of Horticultural Sciences; Gesellschaft für Arzneipflanzenforschung; GDCh

Awards:
CVE1 Medal; Medal of Societas Pharmaceutica Hungarica; Medal of University of Lublin

---

OHMACHT, Robert, Dr.
Institute of Medical Chemistry
University Medical School
Szigeti, 12
7601 Pecs
Phone: +36 72 324122
Fax: +36 72 326373
E-Mail: ohmacht@apacs.pote.hu

Private address:
Park Str. 2
7634 Pecs

Domains of application:
Chemistry of carotenoids; separation sciences (liquid chromatography)

Applied methods:
HPLC; spectroscopic methods

ORSI, Ferenc, Dr. Sci.
Department of Biochemistry and Food Technology
Technical University of Budapest
Muegyetem RDP, 3
1521 Budapest
Phone: +36 1 4632283
Fax: +36 1 4633855
E-Mail: orsi@ch.bmf.hu

Private address:
Romer F. U. 46
1024 Budapest

Domains of application:
Food chemistry; food analysis

Applied methods:
Gas chromatography; HPLC

Membership:
International Association for Cereal chemistry

PATTHY, Mikos, Dr.
Department of Biochemistry
Institute for Drug Research
Berlini, 47-49
1045 Budapest
Phone: +36 1 690011
Fax: +36 1 693229

Domains of application:
Pharmaceutical research; study of the mechanism of action of drug candidates (biochemical background)

Applied methods:
Separation techniques; mosthy HPLC

PAULIK, Ferenc, Dr.
Institute for General and Analytical Chemistry
Technical University of Budapest
Gellert Ter. 4
1111 Budapest
Phone: +36 1 4633417

Private address:
Bogar u. 18/B
1022 Budapest

Domains of application:
Chemical analysis

Applied methods:
Thermal analysis; simultaneous TG, DTG, DTA (derivatograph) quasi isothermal measuring techniques;

evolved gas analysis;
thermodilatometry

Membership:
International Confederation for Thermal Analysis and Calorimetry; European Confederation for Thermal Analysis and Calorimetry

Awards:
Mettler-Award of the International Confederation for Thermal Analysis and Calorimetry; Kurnakov Memorial Medal from Thermoanal, Section of Russian Academy of Sciences

---

PETHO, Gabor, Dr.
Human Serum Production and Medicine Manufacturing Co. Ltd.
Tancsics M. u. 82
2101 Godollo
Phone: +36 28 420307
Fax: +36 28 420177
E-Mail: pethog@human.gau.hu

Private address:
Kassai u. 10
1225 Budapest

Domains of application:
Quality control of raw materials, packaging materials and finished products (sterile liquids, solid dosage forms); analytical method development for R&D division

Applied methods:
UV/VIS; HPLC; GC; AAS; potentiometry

Membership:
ACS

POKOL, Gyorgy, Dr. Sci.
Institute of General and Analytical Chemistry
Technical University of Budapest
Szt. Gellert Ter. 4
1521 Budapest
Phone: +36 1 4631593
Fax: +36 1 4634052
E-Mail: pokol@ch.bme.hu

Private address:
Ida u. 39
1162 Budapest

Domains of application:
Analytical chemistry; structural chemistry; solid space chemistry

Applied methods:
Thermal analysis; x-ray powder diffraction; statistical methods in the evaluation of analytical results

Membership:
International Confederation of Thermal Analysis and Calorimetry

---

POSTA, Jozsef, Dr. Sci.
Department of Inorganic and Analytical Chemistry
Kossuth University
Egyetem Ter. 1
4010 Debrecen
Phone: +36 52 316666
Fax: +36 52 310936
E-Mail: postaj@tigris.klte.hu

Private address:
Tarian u. 4, 4032 Debrecen

Domains of application:
Sample introduction in atomic spectrometry; development of hyphenated techniques for element speciation analysis; development of atomic spectrometric methods for environmental chemistry

Applied methods:
FAAS; GFAAS; ICP-AES; HPLC; CE

Awards:
Award of the Minister of Education

PUNGOR, Erno, Prof. Emeritus
Institute for General and Analytical Chemistry
Bay Zoltan Foundation for Applied Research
Kondorfa u. 1
1509 Budapest
Phone: +36 1 4630502
Fax: +36 1 4630503
E-Mail: pungor@bzaka.hu

Private address:
Meredek u. 4
1112 Budapest

Domains of application:
Analytical chemistry; ion selective electrodes; electroanalysis; oscillometry; flame photometry and atomic absorption

---

RAAL, Tamas, Bach. Sci., Dr.
National Institute of Pharmacy
Zrinyi 3
1051 Budapest
Phone: +36 1 1174044
Fax: +36 1 1171462

Domains of application:
Pharmaceutical quality assurance; pharmaceutical analysis

Applied methods:
UV and IR spectroscopy; HPLC; OPLC

Membership:
European Society of Regulatory Affairs; Drug Information Association; European Forum for Good Clinical Practice

Awards:
RAPS "Special Recognition Award"; "Distinguished Pharmacist"

REPASI, Janos, Dr.
ICN Alkaloida Hungary Co. Ltd.
Kabay J. 29-31
4440 Tiszavasvari
Phone: +36 42 373903
Fax: +36 42 373903
E-Mail: icnalk@szabinet.hu

Domains of application:
Pharmaceutical analysis

Applied methods:
HPLC; GC; TLC; AAS; GC-MS; LC-MS; IR; UV; DSC

Membership:
European Organization for Quality

Awards:
Novicardin Award of Hungarian Academy of Sciences; Honoured Inventor of Hungarian Patent Office; John Kabay Award of ICN Alkaloida

---

SALGO, Andras, Prof. Dr.
Department of Biochemistry and Food Technology
Technical University of Budapest
Muegyetem RKP 3
1111 Budapest
Phone: +36 1 4633854
Fax: +36 1 4633855
E-Mail: salgo@ch.bme.hu

Domains of application:
Molecular biology; food analysis; technology development

Applied methods:
Chromatography; electrophoresis; proximate analysis; flow injection analysis; PCR; immunochemical methods; near infrared spectroscopy; sensory analysis; enzymatic methods
Membership:
ISO

SARKADI, Simon, Dr.
Department of Biochemistry and Food Technology
Technical University of Budapest

Muegyetem RKP 3
1521 Budapest
Phone: +36 1 4633862
Fax: +36 1 4633855
E-Mail: sarkadi@ch.bme.hu

Domains of application:
Teaching biochemistry; food technology; food analysis; research: food analysis, biochemical and molecular biological research, technological research

Applied methods:
Chemical and biochemical analysis of food and feed using different techniques

Membership:
Hungarian Association of Chemistry; Hungarian Association of Biochemistry; FECS

SIMON, Judith, Dr.
Institute for General and Analytical Chemistry
Technical University of Budapest
Szt. Gellert Ter. 4
1521 Budapest

Private address:
Amerikan ut. 52, 1145 Budapest

Domains of application:
Thermal analysis

Applied methods:
Solid state reactions; flammability

Membership:
ICTAC

SZALONTAI, Gabor, Dr. Sci.
University of Veszprem
Egyetem u. 10
Veszprem
Phone: +36 88 422022
E-Mail: unmr1@sparch.mars.vein.hu

Domains of application:
Chemistry, stereochemistry; organomedallic chemistry; spectroscopy

Applied methods:
NMR spectroscopy

SZANTAY, Csaba, Dr.
Gedeon Richter Ltd.
Diosgyori, 21
1475 Budapest
Phone: +36 1 4314952
Fax: +36 1 2605118
E-Mail: cs.szantay@richter.hu

Domains of application:
Structural investigation of synthetic organic molecules

Applied methods:
High resolution NMR

Membership:
Editorial Board of the Journal of Concepts in Magnetic Resonance

---

SZEPESVARY, Pai, Dr. Sci.
Department of Chemistry
Eotvos University
PO Box 32
1518 Budapest
Phone: +36 1 2091173
Fax: +36 1 2090602
E-Mail: szpel@ludens.elte.hu

Private address:
Seroksari ut 38-40
1095 Budapest

Domains of application:
Physical chemistry; analytical chemistry; chemometrics
Applied methods:
Whole instrumental analysis

Awards:
Academic Award, 1970

SZEPESY, Laszlo, Dr. Sci., Dr. techn.
Department of Chemical Technology
Technical University of Budapest
Budafoki, 8
1521 Budapest
Phone: +36 1 4631020
Fax: +36 1 4631913

Private address:
Zenta, 5
1111 Budapest

Domains of application:
Separation techniques; instrumental analysis; biopolymers; environmental analysis

Applied methods:
GC; HPLC; GC-MS; SPE; adsorption

Membership:
Chromatographic Society

SZILAGYI, Laszlo, Dr. Sci.
L. Kossuth University
Egyetem Ter. 1
4010 Debrecen
Phone: +36 52 316666
Fax: +36 52 310936
E-Mail: lszilagyi@tigris.klte.hu

Domains of application:
Carbohydrates; antibiotics (aminoglycosides, glycopeptides, macrolides); oligopeptides; proteins

Applied methods:
NMR spectroscopy

SZOLLOSY, Aron, Dr.
Institute for General and Analytical Chemistry
Technical University of Budapest
Gellert Ter. 4
1111 Budapest
Phone: +36 1 4632293
Fax: +36 1 4633408
E-Mail: szollosy@ch.bme.hu

Domains of application:
Structure elucidation and conformational analysis of organic and organometallic compounds

Applied methods:
NMR spectroscopy

TAKACS, Mihaly, Dr.
Institute of Pharmaceutical Chemistry
Semmelweis University of Medicine
Hogyes E. U. 9
1092 Budapest
Fax: +36 1 2170891

Domains of application:
Pharmaceutical chemistry; pharmaceutical analysis; photochemistry

Applied methods:
Chromatography; TLC; spectroscopy

TAKACS-NOVAK, Krisztina, Dr. Sci.
Institute of Pharmaceutical Chemistry
Semmelweis University of Medicine
Hogyes E. U. 9
1092 Budapest
Phone: +36 1 2171222
Fax: +36 1 2170891
E-Mail: novkri@hogyes.sote.hu

Domains of application:
Pharmaceutical chemistry; pharmaceutical analysis; medical chemistry (QSAR), education and research

Applied methods:
Instrumental analytical methods: electrochemistry, chromatography, spectroscopy; mainly: potentiometry, UV, HPLC, TLC

Hungary

TORK, Ilona, Dr. Chem.
National Institute of Pharmacy
Zrinyi 3
1372 Budapest
Phone: +36 1 1178421
Fax: +36 1 1181168

Private address:
Margareth str. 11
1126 Budapest

Domains of application:
National Institute of Pharmacy; quality control division

TOTH, Gabor, Dr. Sci.
Technical Analytical Research Group of the Hungarian Academy of Sciences
Technical University of Budapest
Szent Gellert Ter. 4
1111 Budapest
Phone: +36 1 4633411
Fax: +36 1 4633408
E-Mail: g-toth@ch.bme.hu

Private address:
Veves P. u. 74
1163 Budapest

Domains of application:
Structure elucidation of synthetic and natural products by different spectroscopic methods; the main fields in the NMR spectroscopy, heterocyclic chemistry and triterpenes; furthermore colixarenes and their host-guest interactions

Applied methods:
Modern 1D and 2D NMR methods: high field (500 MHz) NMR in liquid phase, molecular modelling

Awards:
Szecheny Prof. Award, 1997

TOTH, Klara, Prof. Dr.
Institute for General and Analytical Chemistry
Technical University of Budapest
Szt. Gellert Ter. 4
1111 Budapest
Phone: +36 1 4632273
Fax: +36 1 4633408
E-Mail: toth@ch.bme.hu

Domains of application:
Lecturing on analytical chemistry: on pharmaceutical analysis; supervising graduate and PhD students

Applied methods:
Electroanalytical chemistry; optical techniques; scanning electrochemical microscopy; electrochemical quartz crystal microbalance

Membership:
IUPAC; DAC (FECS)

Awards:
Award given by the Hungarian Academy of Sciences

TYIHAK, Erno, Dr. Sci.
Plant Protection Institute
Hungarian Academy of Sciences
Herman Otto u. 15
1525 Budapest
Phone: +36 1 1558722
Fax: +36 1 1563698

Private address:
Zapor ut. 59
1032 Budapest

Domains of application:
Study of phases of stress syndrome in relation of formaldehyde cycle; biochemical immunization of plants; chemotaxonomical study of stress and resistance factors; analysis of formaldehyde and methylated compounds in biological systems

Applied methods:
UV spectroscopy; NMR

spectroscopy; MALDI MS; thin-layer chromatography; HPTLC; OPLC; HPLC

Awards:
MTESZ Plaquette

VARADI, Maria, Dr.
Central Food Research Institute
Herman Otto u. 15
1536 Budapest
Phone: +36 1 1558982
Fax: +36 1 1558991
E-Mail: h9741var@ella.hu

Private address:
Lukacs Gy. U. 4
1039 Budapest

Domains of application:
Analysis of food

Applied methods:
Near infrared reflectance/transmittance spectroscopy; biosensors

Membership:
AOAC; International Committee for Near Infrared Spectroscopy; FAO/WHO Codex Alimentarius Hungarian Committee; European Organization for Quality

---

VARHEGYI, Gabor, Dr. Sci.
Research Laboratory for Inorganic Chemistry
Hungarian Academy of Sciences
Budaorsi ut 45
1518 Budapest
Phone: +36 1 3193119
Fax: +36 1 3193142
E-Mail: varhegyi@sparc.core.hu

Private address:
Eper u. 2A
1112 Budapest

Domains of application:
Domains of applications in my department; biomass research; characterization of solid fuels; thermal decomposition of organic polymers

Applied methods:
Thermogravimetry-mass spectrometry; analytical pyrolysis with GC-MS

VERESS, Gabor E., Dr. Sci., Prof.
Department of Information Systems
University of Veszprem
Egyetem u. 2
8200 Veszprem
Phone: +36 88 423466
Fax: +36 88 423466
E-Mail: veress@dis.vein.hu

Domains of application:
Development of information systems; metrology; quality management; technical regulation; chemometrics

Applied methods:
Systems theory; mathematical modelling; mathematical statistics; pattern recognition; quality management methods

Membership:
Mathematical Programming Society; Chemometric Society; European Organization for Quality; Regulatory Professionals Society

---

ZARAY, Gyula, Dr. Sci.
Department of Chemical Technology and Environmental Chemistry
Eotvos University
Pazmany Peter Stny. 2
1518 Budapest
Phone: +36 1 2090597
Fax: +36 1 2090602
E-Mail: zaray@ludens.elte.hu

Private address:
Gador u. 136
1222 Budapest

Domains of application:
Environmental analysis: elemental analysis of airborn dust and aerosol samples, speciation of heavy elements; analysis of plant materials studying the response effects of plants to environmental stress; elemental analysis of soils and sediments

Applied methods:
Inductively coupled plasma atomic emission spectrometry (ICP-AES); graphit furnace atomic absorption spectrometry; total reflection x-ray fluorescence spectrometry; HPLC; HPLC/ICP-AES; ICP-AES; ion chromatography

Ireland

SMYTH, Malcolm R., Bach. Sci., Dr. Sci.
Dublin City University
Dublin 9
Phone: +353 1 7045308
Fax: +353 1 7045032
E-Mail: smythm@ccmail.dcv.ie

Domains of application:
Teaching: DCU offers a BSc in analytical science; MSc in instrumental analysis; MSc/PhD programmes in research analytical chemistry; Specialist: electroanalytical chemistry; sensors; instrumentation; separation science; mass spectroscopy

Applied methods:
The university is well equipped with the majority of instrumental methods of analysis (GC, Q-MS/MS, GFAAS, IR, NMR, HPLC, LC-MS, electroanalysis, ESCA, SEM, sensors, thermal gravimetric analysis, diode array spectrometric)

Membership:
Fellow of Institute of Chemistry of Ireland; Fellow of Royal Society of Chemistry

Awards:
Royal Society of Chemistry Award for Chemical Analysis and Instrumentation, 1995

Israel

BERNSTEIN, Joel, Bach. Sci., M. Sci., Dr.
Department of Chemistry
Ben-Gurion University of the Negev
PO Box 653
64105 Beer Sheva
Phone: +972 7 6461111
Fax: +972 7 6472943
E-Mail: yoel@bgumail.bgu.ec.il

Private address:
Brosh st. 3
84965 Omer

Domains of application:
Analytical and characterization of organic solids; in particular polymorphic pharmaceutical materials

Applied methods:
Hot stage microscopy; differential scanning calorimetry; differential thermal analysis; FTIR; CP/MAS; NMR on solids; x-ray powder diffraction; x-ray single crystal structure determination; crystal growing; thermal gravimetric analysis

Membership:
European Crystallographic Committee; American Crystallographic Association;

American Association for the Advancemente of Science; RSC; ACS

DEVRIES, Mattanjah S., Prof.
Department of Chemistry
The Hebrew University of Jerusalem
93708 Jerusalem
Phone: +972 2 6586109
Fax: +972 2 6525037
E-Mail: devries@fh.huji.ac.il

Domains of application:
Laser desorption; mass spectrometry; multiphoton ionization; large organic molecules in and on surfaces

Applied methods:
Laser desorption; mass spectrometry; photoionization

Membership:
ACS; American Physical Society; Israeli Chemical Society

GRUSAKA, Eli, Dr.
Department of Inorganic and Analytical Chemistry
The Hebrew University of Jerusalem
93708 Jerusalem
Phone: +972 2 6585311
Fax: +972 2 6586201
E-Mail: elija@batata.fh.huji.ac.il

Domains of application:
Separation science; chemometrics

Applied methods:
HPLC; CE

Membership:
ACS

MANDLER, Daniel, Dr.
Department of Inorganic and Analytical Chemistry
The Hebrew University of Jerusalem
91904 Jerusalem
Phone: +972 2 6585831
Fax: +972 2 6585319
E-Mail: mandler@vms.huji.ac.il

Private address:
Gordon st
96669 Jerusalem

Domains of application:
Electrochemistry; analytical and physical; surface science

Applied methods:
Electrochemical methods (voltammetry, amperometry); spectroscopy (FTIR, XPS, ellipsometry); microscopy (atomic force microscopy, scanning electrochemical microscopy)

Membership:
International Society of Electrochemistry; Electrochemical Society; ACS

Awards:
E. Bergman Award for young scientists

PELLY, Ithamar Zvi, Dr.
Ben-Gurion University of the Neger
PO Box 653
84105 Beer Sheva
Phone: +972 7 6461291
Fax: +972 7 6472997
E-Mail: ipelly@bgumail.bgu.ac.il

Private address:
Michael Even-Ari 39
Beer Sheva

Domains of application:
Geology (geochemistry; mineralogy, environmental geochemistry hidrology, engineering geochemistry paleontology, structrual geochemistry)

Applied methods:
Atomic absorption; DTA+TGA; microscopy; IR; gas chromatography; spectrophotometry; x-ray diffraction; ICP; SEM; heating stage microscopy; cathod luminescence

Awards:
Under my supervision: M.Sc Thesis: Annual Prize of the Israeli Council for Fertilizers, 1978; M.Sc. Thesis:: Annual Prize of Ore Engineering Convention, 1980

SCHECHTER, Israel, Dr.
Technion-Israel Institute of Technology
Technion City
32000 Itaira
Phone: +972 4 8292574
Fax: +972 4 8292579
E-Mail: israel@tx.technion.ac.il

Domains of application:
Environmental analytical chemistry; electrochemistry

Applied methods:
Application of lasers to analytical chemistry

Membership:
The Israel Chemical Society; IUPAC

Awards:
Anshel-Rich Innovation Award; Fritz-Itaber Prize

Italy

CAMPANELLA, Luigi, M. Dr.
Chemistry Department
University "La Sapienza"
A. Moro, 5
00185 Rome
Phone: +39 6 4463828
Fax: +39 6 490375
E-Mail: campanella@sci.uniroma1.it

Private address:
Via Cesare Abba, 15
00141 Zip

Domains of application:
Biological, medican and pharmaceutical chemistry; conservation and restoration of cultural heritage; electrochemistry and electrochemical devices; environmental chemistry; food chemistry; materials chemistry; methodologiees

Applied methods:
Voltammetry; sensor and biosensor; potentiometry and amperometry; spectrophotometry; x-ray analysis; thermal methods; electrophoresis; chromatographic techniques

Membership:
AAAS; AOAC; SCI; WSMBMP

Awards:
International Capire Prize for a Creative Future, 1994

CASELLI, Maurizio, Full Prof.
Department of Chemistry
University of Bari
Via Orabona, 4
70126 Bari
Phone: +39 80 5442021
Fax: +39 80 5442129
E-Mail: cscpmc48@area.ba.cnr.it

Domains of application:
Analytical chemistry; environmental chemistry; supramolecular chemistry

Applied methods:
Ionic chromatography; electroanalytical methods

DESIMONI, Elio, Full Prof.
Dipartimento di Fisiologia delle Piante Coltivate e Chimica Agraria
Via Celoria, 2
20133 Milano
Phone: +39 2 26607225
Fax: +39 2 2663057
E-Mail:
desimoni@iniucca.csi.unimi.it

Domains of application:
In the field of analytical chemistry: development of statistical techniques for data analysis, development of methodologies for quantifying analytes in environmental matrices

Applied methods:
Electroanalytical techniques
Awards:
IUPAC

GALLETTI, Guido, "Laurea" in Chemistry
Department of Chemistry
University of Bologna
Via G. Selmi, 2
40126 Bologna
Phone: +39 51 259459
Fax: +39 51 259456
E-Mail: galletti@ciam.unibo.it

Domains of application:
GC, HPLC; mass spectrometry (GC/MS); analytical pyrolysis (PY/GC/MS); lignocellulose; sediments

Applied methods:
Applications of GC, PY/GC/MS and HPLC to agro-food research; environment and conservation of cultural heritage; currently partner in two EU research projects on Agro-Industrial research

Awards:
1995: Best poster VI Sumposium Pharmaceutical Analyses; 1992: Best poster 10th International Sumposium on Analytical Pyrolysis; 1984: National Award "L.A. Bonomelli" for scientific paper on juniper essential oil

---

GENNARO, Maria Carla, Full Prof.
Department of Analytical Chemistry
University of Torino
Via P. Giuria, 5
10125 Torino
Phone: +39 11 6707649
Fax: +39 11 6707626
E-Mail: gennaro@ch.unito.it

Private address:
Via C. Lombroso, 21
10125 Torino

Domains of application:
Environmental chemistry; chemometrics; food chemistry; surface and inter-phase chemistry; cultural heritage; ethnopharmacological chemistry; immunochemistry

Applied methods:
AAS, ICP, HPLC, IC, ITR-HPLC, HRGC, SFC, capillary electrophoresis, mass spectroscopy, reflectance spectroscopy

Membership:
Fellow and chartered chemists of RSC

LOCATELLI, Clinio, Assoc. Prof.
Department of Chemistry
University of Bologna
Via Selmi, 2
40126 Bologna
Phone: +39 51 259534
Fax: +39 51 259456

Private address:
Via Angola, 9

44015 Portomaggiore

Domains of application:
Analytical, organic, inorganic and clinical physic chemistry

Applied methods:
Spectroscopic, voltammetric and chromatographic methods

MALITESTA, Cosimino, "Laurea", PhD
Dipartimento di Chimica
Universita degli Studi di Bari
Via Orabona, 4
70125 Bari
Phone: +39 80 5442019
Fax: +39 80 5442026
E-Mail: malitesta@area.ba.cnr.it

Domains of application:
Surface analysis; polymer modified electrodes; chemometrics; spectroscopies of minerals

Applied methods:
Surface spectroscopies; electroanalytical techniques; EQCM

MASCINI, Marco, Full Prof.
Dipartimento di Sanita Pubblica, Epidemiologia e Chimica Analytica Ambientale
University of Florence
Via Gino Capponi, 9
50121 Firenze
Phone: +39 55 2757274
Fax: +39 55 2476972
E-Mail: mascini@cesitl.unifi.it

Domains of application:
Electroanalytical chemistry; biosensors catalytic and affinity; electrochemical optical piezoelectric biosensors

Applied methods:
Application in medicine, environment and food chemistry

Membership:
European Science Foundation. Programme ABI Chairman Concerted Action: Biosensor Stability (ECC) Steering Committee

Awards:
Fifth International Meeting on Chemical Sensors, 1994

PALMISANO, Francesco, "Laurea"
Dipartimento di Chimica
Universita degli Studi di Bari
Via Orabona, 4
70126 Bari
Phone: +39 80 5442016
Fax: +39 80 5442026
E-Mail: f.palmisano@area.ba.cnr.it

Domains of application:
Electrochemical biosensors; chromatography; mass spectrometry

Applied methods:
Clinical chemistry-biomedicine; environmental chemistry; food chemistry

---

PICCARDI, Giovanni, Full Prof.
Department of Public Health and Environmental Analytical Chemistry
Via Gino Capponi, 9
50121 Florence
Phone: +39 55 2757273
Fax: +39 55 2476972
E-Mail: giopic@cesit1.unifi.it

Applied methods:
Ion chromatographic; FIA; AAS; electrochemical techniques

RESCHIGLIAN, Pierluigi, Dr. Anal. Chem.
Department of Chemistry
University of Bologna
Via Selmi, 2
40126 Bologna
Phone: +39 51 259544

Fax: +39 51 259456
E-Mail: resky@ciam.unibo.it

Private address:
Via XXV Aprile 39, 44100 Ferrara

Domains of application:
The research activities carried out in the Department several field in the main traditional areas of chemistry: analytical, industrial, inorganic, organic and physical chemistry

Applied methods:
AAS; field-flow fractionation, UV-vis

RODA, Aldo, Prof.
Department of Pharmaceutical Science
University of Bologna
Via Belmeloro, 6
40126 Bologna
Phone: +39 51 343398
Fax: +39 51 343398
E-Mail: roda@alma.unibo.it

Private address:
Via Saragozza, 18
40123 Bologna

Domains of application:
Organic chemistry; medicinal chemistry; pharmaceutical analysis; analytical chemistry; pharmaceutical technology; industrial microbiology

Applied methods:
Bio and cheminuminescent analytical methods; biosensors applied in clinical pharmaceutical environmental and food analysis; immunoassays; HPLC; GC; mass spectrometry; electrospray interface; imaging using chemiluminescence; application in gene probe

Membership:
IUPAC: delegate for Italy

SABBATINI, Luigia, "Laurea"
Dipartimento di Chimica
Universita degli Studi
Via Orabona, 4
70126 Bari
Phone: +39 80 5442020
Fax: +39 80 5442026
E-Mail: sabba@area.ba.cnr.it

Private address:
Via Bari, 49

Domains of application:
Surface analysis; biomaterials; archaometry

Applied methods:
Surface spectroscopies; FT-IR; electroanalytical techniques (electrosynthesis of polymer films)

Membership:
SCI;Vice-President of Analytical Chemistry Division; IUPAC member

---

SEEBER, Renato, Full Prof.
Department of Physical and Inorganic Chemistry
University of Bologna
V. le Risorgimento, 4
40136 Bologna
Phone: +39 51 6443257
Fax: +39 51 6443690
E-Mail: seeber@ms.fci.unibo.it

Private address:
Strada Collegarola, 128
41100 Modena

Domains of application:
Theoretical chemistry; spectrometry; coordination chemistry; analytical chemistry

Applied methods:
Electroanalytical chemistry: chemically modified electrodes-microelectrodes; molecular electrochemistry; chemometrics

Membership:
Member of RSC and ACS

TADDIA, Marco, Assoc. Prof.
Department of Chemistry
University of Bologna
Via F. Selmi, 2
40126 Bologna
Phone: +39 51 259544
Fax: +39 51 259456
E-Mail: taddia@ciam01.cineca.it

Domains of application:
Inorganic chemistry; organic chemistry; physical chemistry; analytical chemistry; theoretical chemistry; spectroscopy; electrochemistry; polymers science; materials science

---

TOMASSETTI, Mauro, Dr.
Chemistry Department
University "La Sapienza"
A. Moro, 5
00185 Rome
Phone: +39 6 49913725
Fax: +39 6 490631

Domains of application:
Electrochemical sensors; electrochemical biosensors; thermal analysis (Tb-DTA-DSC-TMA); pharmaceutical and biomedical analysis; characterization, conservation and restoration of cultural heritables; food analysis; environmental analysis

Applied methods:
Thermogravimetry, differential thermal analysis, differential scanning calorimetry, thermal mechanical analysis, electrochemical sensor and biosensor, potentiometry, amperometry, voltammetry, spectrophotometry (vis, UV, IR), chromatographic techniques

Membership:
SCI; SIBiol; AICAT

ZAMBONIN, Pier Giorgio, Prof.
Department of Chemistry
University of Bari
Via Orabona, 4
70126 Bari
Phone: +39 80 5443636
Fax: +39 80 5442026

Domains of application:
Preparation-characterization of electroproduced polymeric materials; sensors and biosensors (basic and applications); mass spectrometric methods (coupled or not with chromatography techniques); electroanalytical chemistry; ESCA, Auger, ISS, FTIR

Membership:
Member of the DAC of FECS and of few Commissions of the "Division" itself; member Presidium of Euroanalysis

Norway

BYE, Ragnar, Dr.
Department of Pharmacy
University of Oslo
PO Box 1068
0316 Oslo
Phone: +47 228 56579
Fax: +47 228 54402

Domains of application:
Development of new AAS methods

Applied methods:
AAS: flame, graphite furnace, hydride generation; electroanalytical methods; decomposition methods

Membership:
RSC

GREIBROKK, Tyge, Cand. Ral.
Department of Chemistry
University of Oslo
PO Box 1033
0315 Oslo
Phone: +47 228 55027
Fax: +47 228 55441
E-Mail:
tyge.greibrokk@kjemi.unio.no

Domains of application:
Development of new separation and detection methods related to capillary columns in HPLC, high temperature liquid chromatography, supercritical fluid chromatography and electrochromatography

---

HOVIND, Haavard, M. Sci.
Norwegian Institute for Water Research
Brekkeveien 19
0411 Oslo
Phone: +47 221 85277
Fax: +47 221 85200
E-Mail: haavard.hovind@niva.no

Private address:
Svensenga 98
0882 Oslo
Norway

Domains of application:
Analysis of water, sludges, sediments and biological materials for trace metals, major constituents, trace organic compounds and several inorganic constituents in efluents from industry. The department also does ecotoxicological and microbiological testing

Applied methods:
FAAS; GFAAS; ICP; ICP-MS; HPLC; GC-MS; ISE; photometric methods

---

JELLUM, Egil, Dr.
Institute of Clinical Biochemistry
Rikshospitalet
University of Oslo
0027 Oslo
Phone: +47 228 67035
Fax: +47 228 67029
E-Mail: egil.jellum@rh.uio.no

Domains of application:
Application to analytical chemistry methods to diagnosis of human disease

Applied methods:
TLC; GC; GC-MS; HPLC; CE; ZD-page; gene technology

Membership:
New York Academy of Science; Scadinavian Electrophoresis Society; International Scientific Committees for Organizing various Chromatography Symposia

Awards:
Tswett Medal (USA, 1983), the Martin Award (UK, 1990); the Medinnova Prize (Norway, 1991); the Dimick Award (USA, 1993)

JENTOFT, Nilj Arne, M. Sci.
Dyno Industrier Aja
Svelleveien
2001 Lillestrom
Phone: +47 638 97642
Fax: +47 638 19050
E-Mail: nils-anne.jentoft@di_li.dc.eu.dynoind.com

Private address:
Eidsvollsgate 21
2000 Lillestrom
Domains of application:
Analytical R&D, QA/QC (organic chemistry); emission testing (wood based products)

Membership:
AOAC; Norwegian Chem. Society

JULSHAMN, Kaace H., Dr.
Institute of Nutrition
University of Bergen
Strandgaten 229
5002 Bergen
Phone: +47 552 38250
Fax: +47 552 38095
E-Mail: spor_kj@nutr.tiskuidir.no

Private address:
Breiviksveien 19
5035 Bergen-Sandvikin

Domains of application:
Nutrition, feed and marine feed resources in aquaculture; nutritional quality of sea food; food chemistry; analytical chemistry related to essential nutrients and inorganic contaminants in foos mairly samples of marine origin

Applied methods:
Flame AAS; ETAAS; FIAS; ICP-MS

Membership:
AOAC

KARLSEN, Jan, Dr.
Department of Pharmaceutics
Faculty of Science
PO Box 1068
0316 Oslo
Phone: +47 228 56594
Fax: +47 228 57494
E-Mail: jan.karlsen.@farmasi.uio.no

Private address:
Hagalivn 3
1342 Jar
Norway

Domains of application:
Natural colours and aroma compounds

Applied methods:
GC; TLC; HPLC; CE

KROGH, Mette, Dr.
Institute of Pharmacy
University of Oslo
PO Box 1068
0316 Oslo
Phone: +47 228 56613
Fax: +47 228 54402
E-Mail: mette.krogh@farmasi.uio.no

Domains of application:
Analysis of drugs in biological samples; automated sample preparation

Applied methods:
Solid-phase microextraction; on-line dialysis; high-performance liquid chromatography; cap gas chromatography; mass spectrometry; capillary electrophoresis

---

LUND, Walter, Dr.
Department of Chemistry
University of Oslo
PO Box 1033
0315 Oslo
Phone: +47 228 55534
Fax: +47 228 55441
E-Mail: walter.lund@kjemi.uio.no

Domains of application:
Quantitative determination of chemical elements; ICP-MS; ICP-AES; GFAAS; speciation analysis; environmental chemistry

Applied methods:
ICP-MS; ETV-ICP-MS; ICP-AES; GFAAS; hydride generation; HPLC/CE-atomic spectrometry

Membership:
IUPAC; FECS-DAC

LUNDANES, Elsa, Dr.
Department of Chemistry
University of Oslo
PO Box 1033
0315 Oslo

Phone: +47 228 55553
Fax: +47 228 55441
E-Mail: elsa.lundanes@kjemi.uio.no

Private address:
Oberst Rodes vei 59
1165 Oslo
Norway

Domains of application:
Environmental; bioligically active compounds; pharmaceutical compounds

Applied methods:
HPLC, HTLC; SFC; GC; CE; electrochromatography; MS

PEDERSEN-BJERGAARD, Stig, Dr.
School of Pharmacy
University of Oslo
PO Box 1068
0316 Oslo
Phone: +47 228 56576
Fax: +47 228 54402
E-Mail: stig-pedersen-bjergaard@farmasi.uio.no

Private address:
Solligrenda 82
0491 Oslo
Norway

Domains of application:
Pharmaceutical bioanalysis; pharmaceutical quality control; forensic analysis; instrumental development

Applied methods:
CE; microextraction; chromatography-atomic spectroscopy; mass spectrometry

---

RASMUSSEN, Knut E., Dr.
Department of Drug Analysis
Institute of Pharmacy
PO Box 1068
0316 Oslo
Phone: +47 228 56584
Fax: +47 228 54402
E-Mail:
k.e.rasmussen@farmasi.uio.no

Private address:
Gullhellaun 1
1370 Asker

Domains of application:
Analysis of drugs in biological fluids; pharmaceutical QC; forensic analysis

Applied methods:
HPLC; GC; CE; GC/MS; SPE; dialysis; SPME and other micro-extraction methods

---

ROVSET, Oddvar, Dr.
Norwegian Institute for Air Research
PO Box 100
2007 Kjeller
Phone: +47 638 98240
Fax: +47 638 98050
E-Mail: oddvar.royset@nilu.no

Private address:
Ole Wingers vet 16
0950 Oslo
Norway

Domains of application:
Environmental analysis related to air research (gases, particles, precipitation), anions, cations, major components, heavy metals, mercury; environmental analysis-general (bioligcal material, soils, water)

Applied methods:
ICP-MS; IC; CV-AFS; sold vapour atomic fluorescence spectrometry; CFS, continuous flow analysis; LIMS

SALBU, Brit, Dr.
Laboratory for Analytical Chemistry

Agricultural University of Norway
PO Box 5026
1432 AAS
Phone: +47 679 48350
Fax: +47 649 48359
E-Mail: brit.salbu@lak.nlh.no

Domains of application:
Analytical chemistry; radiochemistry; electron microscopy

Applied methods:
ICP; GFAAS; FIA; ICP-MS; IC;; fibre; fat; carbohydrates; NC; H; S; alpha (pu); beta (90-Sr); gamma; neutron activation analysis; SEM; TEM with XRMA

Awards:
Machmann Legacy

THOMASSEN, Yngvar, Cand. Real
National Instiute of Occupational Health
Gydas vei 8
0033 Oslo
Phone: +47 231 95320
Fax: +47 231 95206
E-Mail: ynguar.thomassen@stami.no

Private address:
Grefoenhollum 1219
0490 Oslo
Norway

Domains of application:
Environmental/occupational exposure: air measurements of a variety of contaminants; bio-monitoring: xenobiotics in blood-urine, tissue

Applied methods:
FAAS; ETAAS; CV-AAS; HGAAS; ICP-AES; GC; GC-MS; HPLC; IC; FT-IR; SEM; x-ray diffraction

---

WIBETOE, Grethe, Dr.
Department of Chemistry
University of Oslo
PO Box 1033
0315 Oslo
Phone: +47 228 55516
Fax: +47 228 55441
E-Mail:
grethe.wibetow@kjemi.uio.no

Domains of application:
Methods for quantitative determination of chemical elements, with emphasis on trace elements

Applied methods:
Molecular absorption spectrometry; atomic absorption spectrometry; flame; graphite furnace; hydride generation; inductively coupled plasma atomic emission spectrometry; inductively coupled plasma mass spectrometry; with graphite furnace

---

WICKSTROM, Torild, Dr.
Norwegian Forest Research Institute
Hogskoleveien 12
1432 AS
Phone: +47 649 49163
Fax: +47 649 42980
E-Mail:
torild.wickstrom@nisk.nlh.no

Private address:
Langretta 20
1279 Oslo

Domains of application:
Forest ecology
Applied methods:
Atomic emission (ICP-AES); ion chromatographic (IC); flow injection analysis (FIA); pH; conductively; titration; TOC

BARANOWSKA, Irena, Dr.
Department of Analytical and General Chemistry
Silesian Technical University
M. Strzody 9
44100 Gliwice
Phone: +48 32 371816
Fax: +48 32 371205
E-Mail:
baranows@zeus.polsl.gliwice.pl

Private address:
ul. Polna 70/20
41600 Swietochlowice

Domains of application:
Analytical methods elaborations using chromatographic methods (HPLC, TLC), spectrophotometry UV-vis; derivative spectrophotometry; atomic absorption spectrometry; voltamperometric methods

Applied methods:
Environmental Analysis /phenols; aromatic and heterocyclic amines; heavy metals; cyanides/; analysis of biological materials /human tissues and fluids - alkaloids; katecholamines; prostaglandines; vitamines; amino acids; analysis of industrial wastes

---

BILEWICZ, Renata, Dr.
Department of Chemistry
University of Warsaw
Pasteura 1
02093 Warsaw
Phone: +48 22 220211
Fax: +48 22 225996
E-Mail: bilewieze@chem.uni.edu.pl

Private address:
Kiochnialma 3
00577 Warsaw

Domains of application:
Bioanalytical chemistry; bioinorganic and biomimetic chemistry; electrocatalytic applications of macrocyclic complexes

Applied methods:
Electrochemistry; Langmuir-Blodgett technique; self-assembly for analytical purposes

Membership:
International Society of Electrochemistry; Polish Chemical Society

Awards:
Polish Awards only

BIZIUK, Marek, Dr. Sci.
Chemical Faculty
Technical University of Gdansk
G. Narutowicz 11/12
80952 Gdansk
Phone: +48 58 471783
Fax: +48 58 472654

Private address:
Gorpody 14B/29
80340 Gdansk

Domains of application:
Environmental analytical chemistry; trace analysis; elemental microanalysis

Applied methods:
Chromatography; spectrophotometric methods; elemental microanalysis

Awards:
Ministery of National Education

BORAWSKA, Maria Halina, Pharm. Dr.
Laboratory of Bromatology
Medical Academy of Bialystok

Kilinskiego 1
15230 Bialystok
Phone: +48 85 422021
Fax: +48 85 424907

Domains of application:
Food safety; food nutrition and health

Applied methods:
Total fiber (based on the method AOAC-1990); analysis of nitrate and nitrite in foodstuffs (based on the reduction of the nitrate ion to the nitrite ion by the use of spongy cadmium(; determination of lead, cadmium, copper and zinc in food, blood and hair

Awards:
3 x Team Award of Secretary of Polish Academy of Sciences

BORKOWSKA-BURNECKA, Jolania, Dr.
Department of Chemistry
Technical University of Wroclaw
Wibrzeze Wispianskiego 87
50370 Wroclaw
Phone: +48 71 3202484
Fax: +48 71 224330
E-Mail:
burnecka@ichn.ch.pwr.wroc.pl

Private address:
Chabrowa 10
53224 Wroclaw

Domains of application:
Inorganic and analytical chemistry

Applied methods:
Spectroscopic methods (emission and absorption); radiochemical; electrochemical methods

---

BULSKA, Ewa, Dr.
Department of Chemistry
University of Warsaw
Pasteura 1
02093 Warsaw
Phone: +48 22 220211
Fax: +48 22 225996
E-Mail: ebulska@chem.uw.edu.pl

Private address:
Afrykaviska 10
03946 Warsaw

Domains of application:
Trace analysis, speciation; sample preparation for environmental and clinical samples; preconcentration techniques for atomic spectrometry

Applied methods:
Atomic absorption and emission spectrometry; gas chromatography and HPLC for speciation; sample preparation-microwave mineralization; flow injection (FI) for atomic spectrometry

Awards:
Post-Doctors Max-Planck Stipendium

BUSZEWSKI, Boguslaw, Dr.
Department of Environmental Chemistry
Nicolaus Copernicus University
Gagarin 1
87100 Torun
Phone: +48 56 14308
Fax: +48 56 542477
E-Mail:
bbusz@anna.mat.uni.torun.pl

Private address:
J. Wibicki str. 100A/41
87100 Torun

Domains of application:
Control and monitoring inorganic and organic hazardous compounds; theory and practice of chromatography and related techniques; sample preparation techniques (absorption extraction

mineralization); utilization and neutralization of wastes

Applied methods:
Absorption; chromatography: TLC, HPLC, GC, GC/MS, CZE; electrochemistry: ion selective electrodes, coulometry; spectroscopy: UV-vis, AAS; sample preparation techniques

Membership:
Chromatographic Society

CHWASTOWSKA, Jadwiga, Dr.
Institute of Nuclear Chemistry and Technology
Dorodna 16
03195 Warsaw
Phone: +48 22 112737
Fax:
E-Mail:
jchuasto@orange.ichtj.waw.pl

Private address:
ul. Kipiow 8
04324 Warsaw

Domains of application:
Analytical chemistry

Applied methods:
Classic methods; spectrophotometry; neutron activity; AAS; separation methods

---

CIESIELSKI, Witold, Dr.
Department of Instrumental Analysis
University of Lodz
W. Linoleya 3
90131 Lodz
Phone: +48 42 355808
Fax: +48 42 783958
E-Mail: zai@chemul.uni.lodz.pl

Private address:
ul. Zgierska 233

Applied methods:
Coulometry; potentiometry; spectrophotometry; polarography; voltammetry

---

CYGANSKI, Andrzej, Prof.
Institute of General and Ecological Chemistry
Technical University of Lodz
ul. Zwirki 36
90924 Lodz
Phone: +48 42 362329
Fax: +48 42 313103

Private address:
ul. Jaracza 92
90244 Lodz

Domains of application:
Analytical and inorganic chemistry; adsorption and catalysis; structural x-ray crystallography and crystallochemistry; chemical technology and environmental protection

Applied methods:
Thermal analysis; thermoanalytical methods; spectrophotometric methods; environmental analysis

Awards:
Cavalier of the Cross Poland's Restitution Order, 1980; Medal of the National Education Commission, 1983

CZERWIECKI, Ludwik, Assoc. Prof.
Department of Food Analysis
Institute of Biotechnology of the Agricultural and Food Industry
Rakowiecka 36
02532 Warsaw
Phone: +48 22 6063871
Fax: +48 22 490426
E-Mail: ibiopros@warman.com.pl

Poland

Domains of application:
Analysis of food contaminants and additives, development of analytical methods

Applied methods:
HPLC; TLC; GC; UV-VIS; AAS

Membership:
IUPAC

Awards:
Award of Scientifical Council of Ministry of Health; Award of Minister of Agriculture

---

DOBKOWSKI, Zbigniew, Dr. Sci.
Department of Analysis and Characterization of Materials
Industrial Chemistry Research Institute
Rydygiera 8
01793 Warsaw
Phone: +48 22 6338298
Fax: +48 22 6338295
E-Mail: zbigdo@aquila.ichp.waw.pl

Private address:
Grojecka 19/25, 02021 Warsaw

Domains of application:
Analysis and characterization polymer materials; environmental protection analysis; research and development (chemical technologies); quality assurance

Applied methods:
GC; GC-MS; LC; GPC (SEC); UV/VIS; FTIR; NMR; microscopy; thermal analysis; TG; DSC; rheology; rheospectrometry; viscometry; mechanical properties testing methods

Membership:
New York Academy of Science; ACS; Polymer Processing Society; EURACHEM; IUPAC

DOBROWOLSKI, Ryszard, Dr.
Faculty of Chemistry
Maria Curie Sklodowska University
Maria Curie Sklodowska Square 3
20031 Lublin
Phone: +48 81 5375784
Fax: +48 81 533669
E-Mail:
rdobrow@hermes.umcs.lublin.pl

Private address:
Radosci 1/7
20530 Lublin

Domains of application:
Developed slurry sampling graphite atomic absorption spectrometry (GFAAS) for speciation of elements; environmental agricultural and industrial analysis; enrichment of trace metals from dilute aqueous solution by activated carbons

Applied methods:
Atomic absorption spectrometry; electrochemical methods; XRF; MIP

Membership:
The Polish Chemical Society

Awards:
Polish Ministry of Education for PhD Thesis

DOJLIDO, Jan, Prof., Dr.
Technical University of Radom
Chrobrego 27
26600 Radom
Phone: +48 22 40031
Fax: +48 22 46420
Private address:
Poli Gojawiczyviskiej 15
01773 Warsaw

Domains of application:
Analysis of water and wastewater

Applied methods:
AAS; GC; HPLC; FIA; fluorimetry

DUDA, Giazyra, Dr.
Karon Marcinkowski University of Medical Sciences
Kakowa 1/2
51878 Poznan
Phone: +48 61 524180
Fax: +48 61 524180

Private address:
Keplera 6A/24
60158 Poznan

Domains of application:
Nutritional food value; human nutrition; civilization diseases

Applied methods:
GLC; TLC; spectrophotometric methods; questionnaire study

DUMKIEWICZ, Ryszard Aleksander, Dr.
Department of Analytical Chemistry and Instrumental Analysis
Maria Curie Sklodowska University
Maria Curie Sklodowska Square 3
20031 Lublin
Phone: +48 81 5375584

Private address:
Zachodimia 2/50
20620 Lublin

Domains of application:
Electrical layer; mechanism of origins of membrane potential; performance characteristic of ion selective electrodes

Applied methods:
Voltammetry; potentiometry; ionometry; spectrometry; AAS

DYBCZYNSKI, Rajmund, Dr.
Institute of Nuclear Chemistry and Technology
Dorodna 16
03195 Warsaw
Phone: +48 22 112737
Fax: +48 22 111532
E-Mail:
rdybczyn@orange.ichtj.waw.pl

Domains of application:
Inorganic trace analysis; preparation and certification of reference materials

Applied methods:
Neutron activation analysis; ion exchange and extraction chromatography; ion chromatography

Awards:
Council for Peaceful Uses of Nuclear Energy (individual: 1963, collective: 1968); Minister of Science Higher Education and Technology (collective: 1983)

FIJALEK, Zbigniew, Dr.
Pharmaceutical Chemistry Department
Drug Institute
Chelmska str. 30/34
00725 Warsaw
Phone: +48 22 412940
Fax: +48 22 410652

Domains of application:
Drug quality and safety; the scope of activity includes drug registration, quality control and research

Applied methods:
Electrochemical methods; EQCM; ASA; spectrophotometry UV/VIS; HPLC

Membership:
Drug Information Association; European Pharcacopeis Commission

---

GENOWEFA, Miszipal, Prof.
Department of Medicinal Chemistry
Medical Academy
Choolzki 6

20093 Lublin
Phone: +48 81 7425174
Fax: +48 81 7410321

Private address:
str. Beskiolzka 45
20869 Lublin

Domains of application:
Teaching of students of medical and analytical chemistry; researching scientific

Applied methods:
Electrochemical; spectrophotometric UV-VIS and IR; HPLC; TLC

Membership:
New York Academy of Sciences

Awards:
2 Minister of Health and Social Helfare

GLAB, Stanislaw, Dr. Sci.
Department of Chemistry
University of Warsaw
Pasteura 1
02093 Warsaw
Phone: +48 22 220975
Fax: +48 22 225996
E-Mail: stanglab@chem.uw.edu.pl

Private address:
Kinowa 25
04030 Warsaw

Domains of application:
Chemistry (analytical, inorganic, organic, physical,electrochemical and spectroscopic methods; chemical sensors and biosensors

Applied methods:
Potentiometry; coulometry; spectrophotometry

Membership:
IUPAC

Awards:
Award of Rettor; Awards of Ministery of National Education

GOCLESKA-ZYLKIEWICZ, Beata, Dr.
Institute of Chemistry
Warsaw University
Al. Pilsudskiego 11/4
15443 Bialystok
Phone: +48 85 457588
Fax: +48 85 457581
E-Mail:
bgodlew@cksr.ac.bialystok.pl

Private address:
Komarowa 11
15682 Bialystok

Domains of application:
Matrix modification and studies of interferences in GFAAS; determination of trace metals in body fluids; different analytical techniques for the study of metal speciation in biological and environmental analysis

Applied methods:
Flame and electrothermal atomic absorption spectroscopy; spectrophotometry; ultrafiltration; liquid chromatography

---

GOLIMOWSKI, Jerzy, Dr.
Department of Chemistry
University of Warsaw
Pasteura 1
02093 Warsaw
Phone: +48 22 220211
Fax: +48 22 225996
E-Mail: golim@chem.uw.ede.pl

Private address:
Lasek Brzozowy 10/42
02792 Warsaw

Domains of application:
Electroanalytical methods;

environmental analytical chemistry

Applied methods:
Voltammetry, AAS; biomonitoring

Membership:
Kolloid Gesellschaft (Germany)

GORECKA, Helena, Dr.
Institute of Inorganic Technology and Mineral Fertilizer
Technical University of Wroclaw
Smoluchowskiego 25
50372 Wroclaw
Phone: +48 71 3202486
Fax: +48 71 3203469

Private address:
Sklodowskiej 15/45
50381 Wroclaw

Domains of application:
Chemistry for agriculture - the new method of mineral; organic-mineral fertilizer production; feed phosphate; fertilizer for horticulture; heavy metals in the environment; trace elements in soil, water, mineral, waste, sewage, sludge

Applied methods:
ICP-OES with application of ultrasonic nebulizer; AAS; ionselective electrode; decompostion of environmental and biological samples using microwave

---

GREGORIWICZ, Zbigniew, Dr. Sci.
Analytical and General Chemistry
Silesian Technical University
Katowicka 5
44100 Gliwice
Phone: +48 371205
Fax: +48 371205

Private address:
Karolinki 14, 44100 Gliwice

Domains of application:
New analytical reaction and reagents

Applied methods:
Industrial chemical analysis

HEFTER, Jacek, Dr.
Analytical Chemistry Division
University of Opole
Oleska Str. 48
45951 Opole
Phone: +48 77 545841
Fax: +48 77 538387

Private address:
Harcerska Str. 10
47220 Kedzierzyn-Kozle

Domains of application:
Chemistry

Applied methods:
Basic analytical methods: classical; spectroscopic and chromatographic methods; GC; pyrolysis GC; thermal analysis

---

HERMANN TADENSZ, W., Prof., Dr.
Department of Physical Chemistry
Karon Marcinkowski University
Swiecickiego 6
60781 Poznan
Phone: +48 61 657005
E-Mail:
hermann@eucalyptics.usoms.poznan.pl
Private address:
Keplera 4H/14
60158 Poznan

Domains of application:
Therapeutic drug monitoring; clinical pharmacokinetics; stability of drugs

Applied methods:
TLC; HPLC; GC; UV-VIS; IR

spectrofluorimetry

Awards:
J. Wolfgang Awards of the Pol. Pharm. Assoc.

HIPPE, Zdzislaw, Prof., Dr.
Department of Computer Chemistry
University of Technology
Powstancow Warszawy Ave. 6
35041 Rzeszow
Phone: +48 17 41519
Fax: +48 17 41519
E-Mail: zshippe@prz.reszow.pl

Private address:
Oxxolinskich 5
35328 Rzeszow

Domains of application:
Development, Dissemination and application of complex computer program systems for chemometrics and analytical methods

Applied methods:
Neural nets; pattern recognition; molecular modelling; simulation of chemical reactions; expert systems technology in analytical processes; computer assisted quality assessment

Membership:
FECS; CODATA

Awards:
1964, State Award in Science and Technology; 1972, Award of Ministry of National Education; 1974, Award of Secretary of Polish Academy of Sciences; 1976, 1978, 1979, 1980, 1983, 1992, Award of Ministry of National Education; 1994, Scientific Town Rzeszow

---

HOPKATA, Hanna, Prof.
Department of Medicinal Chemistry
Medical Academy
Chockiki 6
20093 Lublin
Phone: +48 74 10321

Domains of application:
Tachering of students of medical and analytical chemistry; researching scientific

Applied methods:
Electrochemical; spectrophotometric UV/VIS and IR; HPLC and TLC

Awards:
Minister of Health and Social Welfare

HULANICKI, Adam, Prof., Dr.
University of Warsaw
Pasteura 1
02093 Warsaw
Phone: +48 22 221085
Fax: +48 22 225996
E-Mail: ahulan@chem.uw.edu.pl

Private address:
Staropolska 2
05500 Piaseczno

Domains of application:
Environmental analytical chemistry; trace analysis-inorganic; speciation analysis

Applied methods:
Atomic spectroscopy; potentiometric sensors

Membership:
IUPAC
Awards:
Zawidzki Medal (Polish Chem. Society); M. Sklodowska-Curie Prize (Polish Academy of Science)

JANCZYSZYN, Jerzy, Dr.
Faculty of Physics and Nuclear Techniques
University of Mining and Metallurgy
Al. Mickiewicza 30

30059 Krakow
Phone: +48 12 173003
Fax: +48 12 340010
E-Mail:
janczyszyn@novell.ftj.agh.edu.pl

Private address:
Skarbinskiego 10/47
30071 Krakow

Domains of application:
Nuclear power engineering; solid state physics; nuclear physics applications; high energy physics; environmental physics; radioanalytical methods

Awards:
Individual Award of the Minister of Science and High Education. 3rd degree, 1972; Award of the National Council for the Peaceful Use of Nuclear Energy, for a team, 3rd degree, 1972; Award of the President of the Polish Atomic Energy Agency, 1973

JUSKOWIAK, Bernard, Dr.
Laboratory of Instrumental Analysis
Adam Mickiewicz University
Grunwaldzka 6
60780 Poznan
Phone: +48 61 699181
Fax: +48 61 658008
E-Mail: juskowia@chem.amu.edu.pl

Private address:
Sobieskiego 21F/120
60688 Poznan
Domains of application:
Environmental analysis; determination of bioactive substances and detection of impurities in pharmaceuticals; development and improvement of the methods of traces separation and preconcentration

Applied methods:
Chromatography; spectrofluorimetry; atomic and molecular spectroscopy; voltammetry

Awards:
Ministry of Education, Poland, 1984 and 1994

KAMINSKI, Marian, Dr. Sci.
Department of Quality Control
Petroleum Refinery of Golanisk
Sloaska 135
80718 Golanisk
Phone: +48 58 472110
Fax: +48 58 388250

Private address:
ul. Buroynskiego 8F/5
80463 Golanisk

Domains of application:
Petroleum products analysis; environmental analysis; toxicologic analysis; preparative and analytical HPLC

Applied methods:
HPLC; GC; FTIR; XRF; AAS; coulometry; potentiometry; TLC-FID; standard methods of petroleum products analysis; microwave digestion; standard analytical methods

Awards:
3 Minister of Science and Academic Education of Poland

KARPINSKA, Joanna, Dr.
Institute of Chemistry
Warsaw University
Al. J. Pilsudskiego 11/4
15443 Bialystok
Phone: +48 85 457587
Fax: +48 85 457581
E-Mail:
polikarp@cksr.ac.bialystok.pl

Private address:
Rzemieslnicza 21/28
15773 Bialystok

Domains of application:
Investigation of analytical properties of phenothiazine and dibenzoazepines and thioxantene derivatives and their application into analysis; determination of phenothiazines, dibenzoazepines, thioxantenes and ubiquinone in body fluids

Applied methods:
UV-VIS spectrophotometry; flow injection analysis; HPLC

KOJLO, Anatol, Dr.
Institute of Chemistry
Warsaw University
Al. J. Pilsudskiego 11/4
15443 Bialystok
Phone: +48 85 457587
Fax: +48 85 457581
E-Mail: kojlo@cksr.ac.bialystok.pl

Private address:
ul. Mieszka I 17/98
15475 Bialystok

Domains of application:
Analytical control of pharmaceuticals and analysis of their content in body fluids; analytical control of environmental pollutions

Applied methods:
Flow injection methods with different detection techniques as spectrophotometry; spectrofluorimetry; potentiometry; amperometry; voltammetry and atomic absorption spectrometry

---

KOLODZIEJSKI, Waclaw, Dr.
Department of Analytical and Inorganic Chemistry
Warsaw Medical Academy
Banacha 1
02097 Warsaw
Phone: +48 22 234484
Fax: +48 22 231487
E-Mail:
waclaw@pluton.farm.amwaw.edu.pl

Domains of application:
Electrochemistry and nuclear magnetic resonance in pharmacy and medicine

Applied methods:
Polarographic determination of drugs

KONCKI, Robert, Dr.
Department of Chemistry
University of Warsaw
Pasteura 1
02093 Warsaw
Phone: +48 22 220211
Fax: +48 22 225996
E-Mail: rkoncki@chem.uw.edu.pl

Private address:
Turmoncka 17/175
03254 Warsaw

Domains of application:
Chemical sensors and biosensors; analysis of real samples

Applied methods:
Potentiometry; flow injection analysis; enzymatic methods

KONIECZKA, Piotr, Dr.
Chemical Faculty
Technical University of Gdansk
G. Narutowicz 11/12
80952 Gdansk
Phone: +48 58 472110
Fax: +48 58 472694
E-Mail:
kaczor@altis.chem.pg.gda.pl

Private address:
Irangutta street 115A
80226 Gdansk

Domains of application:
Environmental analytical chemistry; calibration

Applied methods:
Chromatography

KOROLCZUK, Mieczystaw, Dr.
Faculty of Chemistry
Maria Curie Sklodowska University
Maria Curie Sklodowska Square 3
20031 Lublin
Phone: +48 81 5375592

Domains of application:
Trace analysis of mercury and chromium and speciation of this elements

Applied methods:
Stripping voltammetry; AAS

KOSCIELNIAK, Pocoet, Dr.
Department of Analytical Chemistry
Jagiellonian University
R. Ingordeno 3
30060 Krakow
Phone: +48 12 336377
Fax: +48 12 340515
E-Mail: koscieln@trurl.ch.uj.edu.pl

Private address:
Rozrycowa str. 20/33
30417 Krakow

Domains of application:
Chemometrics: calibration, examination of the interference effects, designing of experiments, optimization of analytical procedures; preparation of analytical solid standards; application of the flow injection technique to various fields of anal. chem.

Applied methods:
Atomic absorption spectrometry; spectrophotometry UV/VIS; flow injection technique

Awards:
2 Awards of Ministry of National Education; 6 Awards of Rector of Jagiellonian University

---

KOT, Agata, Dr.
Department of Analytical Chemistry
Technical University of Gdansk
Narutowicza 11/12
80952 Gdansk
Phone: +48 58 472405
Fax: +48 58 472694
E-Mail: agata@altis.chem.pg.gda.pl

Private address:
Aleja Jana Pawka I 21A/15
80462 Gdansk

Domains of application:
CGC and HPLC in environmental chemistry; hyphenated techniques; sample preparation techniques (SPE, SPME, microwave extraction, etc.) used for the organic pollutants

Applied methods:
Chromatography; hyphenated techniques

Awards:
1995, Brussels (Belgium) Royal Academy of Science Award; 1995, Gdansk (Poland) Regional Scientific Academy Award

---

KOWALSKA, Teresa, Dr.
Institute of Chemistry
Silesian University
Szkolna Street 9
40006 Katowice
Phone: +48 32 599978
Fax: +48 32 599978
E-Mail: kowalska@tc3.ich.us.edu.pl

Private address:
Cieszynska Street 5
41500 Chorzon

Applied methods:
Separation techniques (TLC, HPLC, GC)

Awards:
Brown Cross of Merit 1979 and 1993; Numerous Awards Granted by Tector of the Silesian University, Katowice, Poland

---

KOWALSKI, Zygmunt, Dr.
Faculty of Materials Science and Ceramics
University of Mining and Metallurgy
Al. Mickiewicza 30
30059 Krakow
Phone: +48 12 172473
Fax: +48 12 341201
E-Mail: zkowal@uci.agh.edu.pl

Private address:
ul. Tyniecka 10B
30323 Krakow

Domains of application:
Analytical chemistry in material science and environmental control; instruments and sensor development and design

Applied methods:
Electroanalytical methods (polarography, voltammetry, ion selective electrodes); atomic absorption spectroscopy (flame, flameless GFAA, Zeeman GFAA, hydride generation technique); atomic emission spectroscopy

Awards:
Awards of the Polish Min. Of Education; Scientific Award of the Polish Acad. Of Science; Nat. Education Commission Medal

---

KOWALZCYK, Jerzy, Dr.
Institute of Inorganic Chemistry and Metallurgy of Rare Elements
Technical University of Wroclaw
Wyb. Wyspianskiego 27
50370 Wroclaw
Phone: +48 71 224330
Fax: +48 71 224330
E-Mail:
kowalczyk@ichn.ch.pwr.wroc.pl

Private address:
Laka Mazurska 13/3
51164 Wroclaw

Domains of application:
Inorganic chemistry; hydrometallurgy and extraction of non-ferrous metals; analytical chemistry; theoretical chemistry; organometallic and coordination chemistry

Applied methods:
ICP-OE; ASA-flame and electrothermal; AE; classical methods

Awards:
1989, Ministry of Higher Education

KRAWCZYNSKI VEL KRAWCZYK, Tadeusz, Dr.
Department of Inorganic and Analytical Chemistry
Laboratory of Flow Analysis and Chromatography
Pasteura 1
02093 Warsaw
Phone: +48 22 220211
Fax: +48 22 223532
E-Mail: tkvk@chem.uw.edu.pl

Private address:
Harcerska 12
05820 Piastow

Domains of application:
Analytical chemistry: flow injection analysis with electrochemical

detection for environmental and clinical analysis (e.g. Biosensors with biological components immobilized in polymer films)

Applied methods:
FIA amperometry and potentiometry

KUBIAK, Wladyslaw W., Dr.
Department of Analytical Chemistry
University of Mining and Metallurgy
ul. Mickiewicza 30
30059 Krakow
Phone: +48 12 341201
Fax: +48 12 341201
E-Mail: kubiak@uci.agh.edu.pl

Private address:
ul. Jozefitow 11
30039 Krakow

Domains of application:
Analytical chemistry in materials science; analytical chemistry in environmental control; electroanalytical sensors; instruments an sensors design and development

Applied methods:
Electroanalytical techniques; flow analysis

Membership:
Polish Academy of Science

LUKASZEWSKI, Zenon, Prof. Dr.
Institute of Chemistry
Technical University of Poznan
Piotrowo 3
60965 Poznan
Phone: +48 61 782786
Fax: +48 61 782571
E-Mail: lukasz@et.put.poznan.pl

Domains of application:
Surfactants and their metabolites in the environment heavy metals (esp. Thallium) in the environment

Applied methods:
Voltammetry; tensammetry; BIAS; CTAS; MBAS; biodegradability tests

Membership:
International Association on Water Quality; IUPAC

MADEJ, Alina, Dr.
Faculty of Chemistry
Jagiellonian University
R. Ingardena 3
30060 Krakow
Phone: +48 12 336377
Fax: +48 12 340515
E-Mail: madej@trurl.ch.uj.edu.pl

Private address:
Mazowiecka 55/9
30019 Krakow

Domains of application:
Chemometrics; optimization of the analytical procedures; preparation and investigation of analytical solid standards; flow injection analysis; mathematical modelling of analytical processes

Applied methods:
Titrimetry (instrumental titration); atomic absorption and emission spectrometry (AAS, AES); spectrophotometry UV/VIS; EPR spectroscopy

Membership:
Polish Chemical Society

MAJ-ZURAWSKA, Magdalena, Dr.
Department of Chemistry
University of Warsaw
Pasteura 1
02093 Warsaw
Phone: +48 22 220211
Fax: +48 22 225996
E-Mail: nmajzur@chem.uw.edu.pl

Poland

Domains of application:
Environmental analytical chemistry; trace analysis inorganic; speciation analysis

Applied methods:
Potentiometry; ISE; electroanalytical methods

MAKUCH, Bogumito, Dr.
Chemical Faculty
Technical University of Gdansk
G. Narutowicz 12/14
80952 Gdansk
Phone: +48 58 472310
Fax: +48 58 472694
E-Mail:
chemmol@oltis.chem.pg.gda.pl

Private address:
Leszczynskich str. 1G/19
80464 Gdanxk

Domains of application:
Environmental analytical chemistry; trace analysis

Applied methods:
Liquid chromatography (HPLC, TLC)

MARCINIEC, Barbara, Prof.
Department of Pharmaceutical Chemistry
K. Marcinskawski University of Medical Sciences
Grurwaldzka 6
60780 Poznan
Phone: +48 61 658635
Fax: +48 61 659568

Private address:
Wilkonskich Str. 13
62020 Skarzedr

Domains of application:
Analysis of drugs (new methods of determination and detection); thermal, photochemical and radiochemical stability of drugs; kinetics and mechanism of decomposition reaction of drugs in solutions and in the solid state

Applied methods:
UV-VIS; IR; MS; GC-MS; TLC; HPLC; ASA; EPR; DTA; DSC

Membership:
IUPAC; International Federation of University Women

MATUSIWEICZ Henryk, Prof.
Institute of Chemistry
Poznan University of Technology
Piotrowo 3
60965 Poznan
Phone: +48 61 782312
Fax: +48 61 782571
E-Mail:
henryk.matusiewicz@fct.put.poznan.pl

Private address:
Os. Stare Zegrze 7/11
61249 Poznan

Domains of application:
Development of atomic spectrometric methods of analysis; application of analytical methods to problems of environmental, clinical chemistry and inorganic trace analysis; sample preparation methods

Applied methods:
Flame photometry; atomic absorption spectrometry; microwave induced plasma atomic spectrometry (emission); microwave assisted sample digestion techniques

Membership:
IUPAC

MICHALSKA, Agata, Dr.
Department of Chemistry

University of Warsaw
Pasteura 1, 02093 Warsaw
Phone: +48 22 220211
Fax: +48 22 225996
E-Mail: agatam@chem.uw.edu.pl

Domains of application:
Environmental analytical chemistry; trace analysis inorganic; speciation analysis

Applied methods:
Potentiometric methods; electroanalytical methods

Awards:
Foundation for Polisk Science, 1994

MICHALSKI, Raymund, Dr.
Institute of Environmental Engineering
Polish University of Science
M.C. Saodowskiej 34
41819 Zabrze
Phone: +48 32 1716481
Fax: +48 32 1717470
E-Mail: ipis@usctoux1.cto.us.edu.pl

Private address:
ul. Gen. W. Sikorskiego 123
41809 Zabrze

Domains of application:
Engineering of environmental

Applied methods:
Gas chromatography (GC-MS, GC-FID); ion chromatography

MIGDALSKI, Jan, Dr.
Faculty of Materials Science and Ceramics
University of Mining and Metallurgy
Al. Mickiewicza 30
30059 Krakow
Phone: +48 12 172474
Fax: +48 12 341201
E-Mail: migdal@uci.agh.edu.pl

Private address:
ul. Roweckiego 25/20
30348 Krakow

Domains of application:
Analytical chemistry in material science and environmental control; instruments and sensor development and design

Applied methods:
Electroanalytical methods (polarography, voltammetry, ion selective electrodes); spectroscopy (UV-VIS, AAS)

Awards:
Awards of the Polisk Min. Of Education Scientific; Award of the Polisk Acad. of Science

---

MULLER, Ewa, Dr.
Optical Emission Laboratory
Institute of Non-Ferrous Metals
Sowinskiego 5
44101 Gliwice
Phone: +48 32 380468
Fax: +48 32 316933

Private address:
ul. Zigbia 13/1
44100 Gliwice

Domains of application:
Analytical method elaboration using optical emission method; investigation of matrix effects; carrier effects; interelement effect; manufacturing of certified reference materials and also standard; reference materials in collaboration with NIST

Applied methods:
Optical emission methods with classical excitation surces and inductively coupled plasma; production of certified reference materials for non-ferrous metals and their alloys using optical emission

method as a base of investigation

NAMIESNIK Jacek, Dr.
Chemical Faculty
Technical University of Gdansk
G. Narutowicz 11/12
80952 Gdansk
Phone: +48 58 422110
Fax: +48 58 472694
E-Mail:
chemanal@altis.chem.pg.gda.pl

Private address:
Posejdon Street 16
80299 Gdansk

Domains of application:
Environmental analytical chemistry; trace analysis
Applied methods:
Chromatography; hyphenated methods

Membership:
IUPAC

Awards:
Ministery of National Education Awards (3 times)

PAKYS, Macin, Dr.
Department of Chemistry
University of Warsaw
Pasteura 1
02093 Warsaw
Phone: +48 22 220211
Fax: +48 22 225996
E-Mail: mpalys@chem.uw.edu.pl

Domains of application:
Electrochemical analysis; electroanalysis

Applied methods:
Voltammetry; stripping technology

PARCZEWSKI, Andrzej, Prof. Dr.
Department of Analytical Chemistry
Jagiellonian University
Ingardena 3
30060 Krakow
Phone: +48 12 336377
Fax: +48 12 340515
E-Mail: parczews@trurl.ch.uj.edu.pl

Private address:
Golaska 17/18
30619 Krakow

Domains of application:
Statistical evaluation and processing of experimental data; calibration of the methods of multicomponent analysis; examination of interference effects; design of experiments and optimization of analytical procedures; mathematical modelling of processes

Applied methods:
instrumental titration; UV-VIS spectrophotometry; potentiometry; AAS; AES; GC; HPLC; TLC

PITAK, Bohdan A., Dr.
Institute of Drug Research
Medical University
Banacha 1
02097 Warsaw
Phone: +48 22 235976
Fax: +48 22 231487

Private address:
Marzanny st. 10
02649 Warsaw

Domains of application:
Drugs analysis

Applied methods:
Classical methods of analysis; spectrophotometry: UV-VIS, IR, TLC, GC, HPLC; analysis of aminoacids and aminosugars; pH metry; potentiometry; conductimetry

Membership:
Polish Pharmaceutical Society; The

University of Texas M.D. Anderson Associates, Houston, USA

PLOTKOWIAK, Zyta, Dr.
Department of Pharmaceutical Chemistry
University of Medical Sciences
Grunwaldzka 6
60780 Poznan
Phone: +48 61 658635
Fax: +48 61 515102

Private address:
Przybuszewski Street 45/4
60356 Poznan

Domains of application:
Analysis of drug; determination of drug stability in aqueous solution and in solid phase; kinetic of drug degradation processes

Applied methods:
Spectroscopic; chromatographic (TLC, HPLC); electrochemical

Awards:
The Gold Cross of Deserts (Consil State, Warsaw, Poland, 1977); Polonia Restitura Cross (Consil State, Warsaw, Poland, 1988); Bieganski Award (K. Marcinkowski Univ. Of Med. Sci. Poznan, Poland, 1989)

POLKOWSKA, Zaneta, Dr.
Chemical Faculty
Technical University of Gdansk
G. Narutowicz 11/12
80952 Gdansk
Phone: +48 58 472110
Fax: +48 58 472694
E-Mail: zaneta@altis.chem.pg.gda.pl

Private address:
Ortowska 7C/24
80347 Gdansk

Domains of application:
Environmental analytical chemistry

Applied methods:
Chromatography

PRZYSLAWSKI, Juliusz, Dr.
Department of Bromatology and Human Nutrition
Karon Marcinkowski University of Medical Sciences
Lakowa 1/2
61878 Poznan
Phone: +48 61 524180
Fax: +49 61 524180

Private address:
Krakowska 5/10
61889 Poznan

Domains of application:
Nutritional valve of food; human nutrition; civilization diseases

Applied methods:
GCC methods; TLC methods; spectrophotometric methods; questionnaire study

---

PSZONICKI, Leon, Dr.
Institute of Nuclear Chemistry and Technology
Dorodna 16
03195 Warsaw
Phone: +48 22 112737
Fax: +48 22 111532
E-Mail:
ipszonic@orange.ichtj.waw.pl

Private address:
Plocka 4/26
01231 Warsaw

Domains of application:
Activation analysis; atomic absorption and atomic emission analysis; spectrophotometry; analytical separation methods; radiochemistry; radiation chemistry;

radiation technology; environment protection technologies and diagnostic methods; nuclear technique

Applied methods:
AES, AAS; spectrofluorescence

PTASZYNSKI Begnan, Dr. Sci.
Institute of General and Ecological Chemistry
Technical University of Lodz
ul. Zwirki 36
90924 Lodz
Phone: +48 42 362329
Fax: +48 42 313103

Private address:
ul. Julianowska 5/7
91473 Lodz

Domains of application:
Analytical and inorganic chemistry; adsorption and catalysis; structural x-ray crystallography and crystallochemistry; chemical technology and environmental protection

Applied methods:
Thermal analysis; thermoanalytical methods; spectrophotometric methods; environmental analysis; x-ray analysis

Awards:
Golden Cross of Merity, 1982

PUZANOWSKA-TARASIEWICZ, Helena, Prof. Dr.
Department of Inorganic and Analytical Chemistry
Institute of Chemistry
Al. Pilsudskiego 11/4
15443 Bialystok
Phone: +48 85 437580
Fax: +48 85 457581
E-Mail:
chemfuw@cksr.ac.bialystok.pl

Private address:
Rowna Str. 15
15336 Bialystok

Domains of application:
Analysis of drugs (mainly phenothiazine derivatives) and biological active substances; determination of trace metals in biological and environmental materials; studies of organic, toxic substances and theral oils in air

Applied methods:
Flow injection; AAS (derivative spectrophotometry); GC

Awards:
Polish Chemistry Society (1974); Polish Ministry of Education (1974, 1982, 1988)

PYRZYNSKA, Krystyna, Dr.
Department of Chemistry
University of Warsaw
Pasteura 1
02093 Warsaw
Phone: +48 22 220211
Fax: +48 22 223532
E-Mail: kryspyrz@chem.uw.edu.pl

Private address:
Wysockiego 20/118
03388 Warsaw

Applied methods:
Solid phase extraction; atomic absorption spectrophotometry; flow injection analysis

---

ROKOSZ, Andrzej, Dr.
Department of Analytical Chemistry
Jagiellonian University
Karasia 3
30060 Krakow
Phone: +48 12 336377
Fax: +48 12 340515
E-Mail: @trurl.ch.uj.edu.pl

Private address:
ul. Imbramowska 3/10
31212 Krakow

Domains of application:
Chemometrics: design of experiments, calibration and optimization of analytical procedures; examination and elimination of analytical interference effects; math. Modelling of analytical processes; preparation and examination of analytical solid standards

Applied methods:
Atomic emission and absorption spectrometries; absorption spectrophotometry UV-VIS; precision titrimetry

Membership:
Polish Chemical Society

SLIWICK, Jozef, Dr.
Institute of Chemistry
Silesian University
Szkolna 9
40006 Katowice
Phone: +48 32 599978
Fax: +48 32 599978

Private address:
Drozdow 13
40530 Katowice

Domains of application:
Chromatographic evalution of hydrophobicity of the biologically active organic ompounds; chromatographic and spectroscopic analysis in evaluation of chemical transformations of environmentally present organic compounds

Applied methods:
Chromatography; IR absorption spectroscopy

Membership:
Deutsche Gesellschaft für Fettwissenschaft; American Microchemical Society

Awards:
Award of the Minister of Education; Awards of the Rector of Silesian University

---

SLUSARSKI, Ludomie, Dr.
Institute of Polymers
Technical University of Lodz
Zeromskiego 116
90538 Lodz
Phone: +48 42 362543
Fax: +48 42 362543
E-Mail: sludpol@lodz1.p.lodz.pl

Domains of application:
Analysis of polymers and components of polymer system (polymer supported catalysts, fillers, modifiers); in particular: structure of polymers surface, networks structure, interface interactions, mechanism of thermal degradation and combustion of polymers

Applied methods:
FTIR-ATR; NMR (H1, C13), WAXS, SAXS; gas chromatography including inverse, contact angle measurement, derivatography, oxygen index measurements, elementary analysis

Membership:
Materials Research Society; European Access Network; IUPAC

Awards:
Ministry of National Education (3) a member of three editorial boards in journals on polymers and material science

---

Poland

SOBKOWIAK, Andrzej, Dr.
Faculty of Chemistry
Rzeszow University of Technology
Aleja Powstaricow Warszawy 6
35959 Rzeszow
Phone: +48 17 549830
Fax: +48 17 43655
E-Mail: asobkow@prz.rzeszow.pl

Applied methods:
Electrochemical methods; gas chromatography; UV-VIS spectroscopy

Membership:
Polish Chemical Society; ACS; Society for Electroanalytical Chemistry; Society of Chemical Industry

Awards:
Scientific Award of the Minister of Education; Scientific Award of the Polish Chemical Society

---

STANKIEWICZ, Wieslaw Kazimierz, Dr.
Institute of Non-Ferrous Metals
Sowinskiego 5
44100 Gliwice
Phone: +48 32 380739
Fax: +48 32 316933

Private address:
ul. Cytadeli Warszawskiej 7/1
44121 Gliwice

Domains of application:
Analytical chemistry of the materials of non-ferrous metals metallurgy; new analytical solutions and developments in instrumental methods of analysis with various excitation sources; XRF; UV-VIS spectrophotometry

Applied methods:
Atomic absorption spectrometry; optical emission spectroscopy; x-ray fluorescence spectroscopy; classical wst chemistry methods; UV-VIS spectrophotometry

Membership:
Worls Working Group of Metrology in Chemistry

STOJEK, Zbihniew, Dr.
Department of Chemistry
University of Warsaw
Pasteura 1
02093 Warsaw
Phone: +48 22 220211
Fax: +48 22 225996
E-Mail: stojek@chem.uw.edu.pl

Private address:
ul. Korfantego 45
04196 Warsaw

Domains of application:
Electroanalysis; surface analysis

Applied methods:
Pulse and linear sweep voltammetries; EDS; stripping analysis

Membership:
ACS; International Society of Electrochemistry

STOLYHNO, Andrzej, Dr. Sci.
Faculty of Chemistry
Technical University of Gdansk
Narutowicza 11/12
80952 Gdansk
Phone: +48 58 471623
Fax: +48 58 478604
E-Mail:
stolyhwo@altis.chem.pg.gda.pl

Private address:
Sosnickiego 1
80307 Gdansk

Domains of application:
Lipids (polar and non-polar); food analysis; food contaminants; food

additives; food legislation (including EN legislation)

Applied methods:
GC (HR-GC); HPLC-DAD; laser light scattering; SFC-DAD; FID, LLSD; CE; spectrophotometry UV-VIS; IR; NMR (500 MHz and 200 MHz)

SYBIRSKA, Halina, Dr.
Department of Forensic Toxicology
Silesian Academy of Medicine in Katowice
Medykow 18
40752 Katowice
Phone: +48 32 1521331
Fax: +48 32 1527591

Private address:
Grazynskiego 5/70
40126 Katowice

Domains of application:
Forensic of medicine and toxicology

Applied methods:
Chromatography: TLC, GC, HPLC, FPIA

SYPRYNOWICZ, Zdzislaw, Prof. Dr.
Department of Chemical Physics and Physico-chemical Methods of Separation
Maria Curie Sklodowska University
Maria Curie Sklodowska Square 3
20031 Lublin
Phone: +48 81 5375501
Fax: +48 81 33669
E-Mail:
suprynow@hermes.umcs.lublin.pl
Private address:
Kraszewskiego 3/5
20431 Lublin

Domains of application:
Environmental analysis; biochemical and clinical analysis; modern sample preparation methods: supercritical fluid extraction, solid phase extraction, supported liquid membranes, accelerated solvent extraction

Applied methods:
Capillary gas chromatography; high performance liquid chromatography; capillary zone electrophoresis; capillary electrochromatography

Membership:
International Association of Environmental Analytical Chemistry

Awards:
Tswett Award, 1994; Officer Cross of Polonia Restitute, 1996

SZMYD, Ewa, Dr.
Department of Analytical Chemistry
Institute of Non-Ferrous Metals
Sowinskiego 5
44100 Gliwice
Phone: +48 32 380278
Fax: +48 32 316933

Private address:
ul. Koscinski 36/5
44100 Gliwice

Domains of application:
Analytical chemistry of the materials of non-ferrous metals metallurgy; new analytical solutions and developments in instrumental methods of analysis with various excitation sources; XRF; UV-VIS spectrometry and others for metallurgical origin materials

Applied methods:
Atomic absorption spectrometry; optical emission spectroscopy; x-ray fluorescence spectroscopy; classical chemistry methods; UV-ViS spectrophotometry

---

Poland

SZYMANSKI, Kazimierz, Dr.
Faculty of Civil Engineering and Environmental Engineering
Technical University of Koszalin
Raclawicka Str. 15/17
75620 Koszhlin
Phone: +48 94 427881
Fax: +48 94 427652
E-Mail:
szymansk@low.tu.koszalin.pl

Private address:
Engelsa str. 47
75361 Koszalin

Domains of application:
Environmental protection; sludge (waste) treatment

Applied methods:
AAS; GC

Membership:
Water Quality International

Awards:
II Degree Award Ministry of Education

TROJANOWICZ, Marek, Dr.
Department of Chemistry
Laboratory for Flow Analysis and Chromatography
Pasteura 1
02093 Warsaw
Phone: +48 22 223532
Fax: +48 22 223532
E-Mail: trojan@chem.uw.edu.pl

Domains of application:
Flow injection analysis for environmental applications; biosensors for chimical and environmental applications; ion-chromatography and capillary electrophoresis for speciation of trace and macro components; automation of chemical analysis

Applied methods:
Potentiometry; amperometry; voltammetry; potentiometric stripping; piezoelectric detection; HPLC; capillary electrophoresis; UV-VIS spectrophotometry; AAS

Awards:
European Community University Professorship to University of Munster, Germany, 1993; Royal Society Visiting Professorship to University of Strathdyde, Glasgow, UK, 1994

TWARDOWSKA, Irena, Dr.
Institute of Environmental Engineering
Polish Academy of Sciences
M. Sklodowska-Curie St. 34
41918 Zabrze
Phone: +48 32 1716481
Fax: +48 32 1717040
E-Mail: ipis@usctoux1.cto.us.edu.pl

Domains of application:
Analysis of industrial wastes, soil, bedrock, pore solutions; interdisciplinary studies on impact assessment of point and nonpoint (large area) sources of contamination on the terrestrial and aquatic environment; pathways, accumulation and release

Applied methods:
AAS; inductively coupled plasma atomic emission spectrometry; FT-IR spectrometry; UV-VIS spectrophotometry; HPLC; ion chromatography

Membership:
International Society of Ecotoxicology and Environmental Safety; Water Environment Federation; International Society for Optical Engineering; Air and Waste Management Association; The Planetary Society

Awards:
Awards by the Scientific Secretary of the Polish Academy of Sciences (1974, 1981, 1989); Awards for the Outstanding Contribution to the Development of Katowice Region, by the Regional Council, Silver Award (1976, Golden Award (1985)

WALAS, Stanislaw, Dr.
Institute of Chemistry
Jagiellonian University
Ingardena 3
30060 Krakow
Phone: +48 12 336377
Fax: +48 12 340515
E-Mail: walas@trurl.un.uj.edu.pl

Domains of application:
Chemometrics; optimization of the analytical procedures; preparation and investigaton of analytical solid standards; flow injection analysis; mathematical modelling

Applied methods:
Atomic absorption and emission spectrometry; flow injection analysis

---

WARDENCHI, Waldemar, Dr.
Department of Analytical Chemistry
Technical University of Gdansk
Narutowicza 11/12
80952 Gdansk
Phone: +48 58 472110
Fax: +48 58 472694
E-Mail:
chemanal@albis.chempg.qda.pl

Private address:
Krsta Str. 14
80217 Gdansk

Domains of application:
Environmental analytical chemistry; trace analysis

Applied methods:
Chromatography; hyphenated methods

WAS, Jolanta, Dr.
Institute of Forensic Research
Westerplatte 9
31033 Kracow
Phone: +48 12 228755
Fax: +48 12 223850
E-Mail: zshebens@cyf-kr.edu.pl

Private address:
Razyckiego 4/89
31324 Cracow

Domains of application:
Microtraces/fibres; paints; glass; soil; toners; polymers analysis

Applied methods:
FTIR microspectroscopy; SEM/EDX; x-ray fluorescence; optical microscopy: stereo-, brightfield-, polarised light; fluorescence; hot stage

Membership:
European Fibers Group

WAWER, Inona, Dr.
Department of Physical Chemistry
Medical University of Warsaw
Banacha 1
02098 Warsaw
Phone: +48 22 236297
Fax: +48 22 231487
E-Mail: wawer@farm.anwaw.edu.pl

Private address:
Barcelonska 3
02762 Warsaw

Applied methods:
NMR; EPR

WITKIEWICZ, Zygfryd, Prof. Dr.
Institute of Chemistry
Military University of Technology
ul. Kaliskiego 2

01489 Warsaw
Phone: +48 22 6859427
Fax: +48 22 6669041

Private address:
Spychwska 11, 01472 Warsaw

Domains of application:
Chromatographic analysis of organic compounds; analysis of toxic substances including environmental pollutions

Applied methods:
GC, HPLC; TLC; overpressured layer chromatography

Membership:
IUPAC; The Chromatographic Society

Awards:
Polish Academy of Sciences; Polish Society of Chemistry; Ministry of Environmental Protection; Russian Society of Chromatography

---

ZAREBA, Stanislaw, Prof.
Department of Medicinal Chemistry
Medical Academy
Chodzki 6
20093 Lublin
Phone: +48 71 10321

Private address:
Mikotaja Kopernika 19

Domains of application:
Teaching of students of medical and analytical chemistry; researching scientific

Applied methods:
Electrochemical; spectrophotometric UV-VIS and IR; HPLC and ILC

Awards:
Minister of Health and Social Welfare (3)

ZIEBA-PALUS, Janina, Dr.
Department of Criminalistic
Institute of Forensic Research
Westerplatte 9
31033 Kracow
Phone: +48 12 228755
Fax: +48 12 223850
E-Mail: zshebens@oyf-kr.edu.pl

Domains of application:
Microtraces/paints; polymers; fibres; soil; inks; toners

Applied methods:
FTIR microspectroscopy; SEM/EDX; x-ray fluorescence; optical microscopy

Membership:
European Paint Group

ZYGMUNT, Bogdan, Dr.
Chemical Faculty
Technical University of Gdansk
G. Narutowicz 11/12
80952 Gdansk
Phone: +48 58 471010
Fax: +48 58 472694
E-Mail:
chemanal@altis.chem.pg.qda.pl

Private address:
Kolberga Str. 23/37
81881 Sopot

Domains of application:
Environmental analytical chemistry; trace analysis
Applied methods:
Chromatography; hyphenated techniques (GC-MS)

Awards:
Ministry of National Education Award (for doctoral thesis)

ZYRNICKI, Wieslaw, Dr.
Chemistry Department
Technical University of Wroclaw

Wyb. Wyspianskiego 27
50370 Wroclaw
Phone: +48 71 3202494
Fax: +48 71 224330
E-Mail: zyrnicki@ichn.ch.pwr.wr.pl

Private address:
Poznanska 19/31
53631 Wroclaw

Domains of application:
Chemistry; biotechnology; technology; chemical engineering

Applied methods:
Analytical chemistry; spectroscopy

Awards:
Awards of Ministry of National Education for Research and Education (1979, 1985, 1988, 1990)

Romania

BAILESCU, George-Emil, Dr. Sci.
Department of Analytical Chemistry
Faculty of Chemistry
Blvd. Republicii 13
70346 Bucharest
Phone: +40 1 4102279
Fax: +40 1 4102279
E-Mail: gebaiu@unichim.sfos.ro

Private address:
Plevnei Street 18
76325 Bucharest

Domains of application:
Analytical chemistry and chemical analysis

Applied methods:
Trace analysis of inorganic compounds; separation techniques; electrometric methods; spectrometric methods

Membership:
DAC-FECS; member in Editorial Board of Analytical Letters

DANET, Andrei, Dr.
Department of Analytical Chemistry
Faculty of Chemistry
Sos. Panduri 90-92
76235 Bucharest
Phone: +40 1 4102279
Fax: +40 1 4102279
E-Mail: danet@unichim.sfos.ro

Domains of application:
Analytical chemistry and chemical analysis

Applied methods:
Flow methods of analysis with chemiluminescence, spectrometric and electrochemical detection; environmental monitoring by using biosensors

---

GOCAN, Simion, Dr.
Department of Analytical Chemistry
Faculty of Chemistry
Arany Janos 11
3400 Cluj-Napoca
Phone: +40 46 190038
Fax: +40 46 190818
E-Mail: sgocan@chem.ubbcluj.ro

Domains of application:
Pharmaceuticals; environmental; water

Applied methods:
GC; TLC; preparative/process chromatography; low-pressure chromatography; SPE

Membership:
The Chromatographic Society; Association of Official Analytical Chemist

---

LUCA, Constantin, Dr. Sci.
Faculty of Industrial Chemistry

Polytechnical University of
Bucharest
Polizu 1
78126 Bucharest
Phone: +40 1 6503289
Fax: +40 1 3111796

Private address:
Str. Marin Serghiescu 11
Bucharest

Domains of application:
Analytical chemistry; instrumental analysis; supramolecular chemistry; metal-ligand interaction; transport processes and carrier design

Applied methods:
Analytical voltammetry; ion-selective electrodes; chemically modified electrodes; synthesis of functional macrocyclic receptors

Membership:
IUPAC; DAC-FECS

Awards:
G. Spacu Prize of Romanien Academy; Scientific Merit Order; Scientific Merit Medal

---

MAGEARU, Vasile, Prof.
Department of Analytical Chemistry
Faculty of Chemistry
Republicii 13
70348 Bucharest
Phone: +40 1 6159249
Fax: +40 1 6159249
E-Mail: vmagearu@unichim.sfos.ro
Private address:
Ariesului 13
Bucharest

Domains of application:
Analytical chemistry; analytical biochemistry

Applied methods:
Electrochemistry; ion-selective electrodes; sensors and biosensors in control of biotechnological processes in clinical analysis

Membership:
Romanian Analytical Chemistry Society

PATROESCU, Constantin, Dr.
Department of Analytical Chemistry
Faculty of Chemistry
Blvd. Republicii 13
70346 Bucharest
Phone: +40 1 4102279
Fax: +40 1 4102279

Private address:
Brasov Street 30
Bucharest

Domains of application:
Analytical chemistry and chemical analysis

Applied methods:
Analysis of roks and minerals; environmental analysis; analysis of geothermal water; classical methods (wet method); spectrometric methods; electrometric methods

---

POPESCU, Alexandru, Dr. Chem.
Department of Analytical Chemistry
University of Craiova
A.I. Cuza 13
1100 Craiova
Phone: +40 51 144621
Fax: +40 51 415077

Private address:
Ulmului Street 15
Craiova

Domains of application:
Sensors for potentiometric determination; the separation of the compounds by GLG; thiocyanato-chrom(III) complexes in analytical chemistry; voltametric study of differently pretreated gold electrodes

Applied methods:
Potentiometry; bipotentiometric; chronopotentiometric; voltametric; GLG

RADU, Gabriel-Lucian, Dr.
Department of Analytical Chemistry
National Institute for Biological Sciences
Spl. Independentei 296
77748 Bucharest
Phone: +40 1 2207909
Fax: +40 1 2207695
E-Mail: lucian@ibd.dbio.ro

Private address:
Aleea Rotunda, 4
74608 Bucharest

Domains of application:
Bioelectrochemistry; biotechnology; control of biomaterials, cosmetics, drugs; environmental analysis

Applied methods:
Electrometrical methods; electronical microscopy (TEM); biosensors; electrophoresis

Membership:
Planetary Society; DAC-FECS

Awards:
Silver Medal 45th World Exhibition of Invention, Research and Industrial Innovation, Brussels Eureka 1998; Silver Medal at 23th Salon International des Inventions, Geneve,
1995

SILVIU, Dumitru-Popescu, Dr.
Institute of Chemical Research
Splaiul Independentei 202
77208 Bucharest
Phone: +40 1 6153299
Fax: +40 1 3123493

Private address:
Vigoniei Street 8
77402 Bucharest

Domains of application:
Analytical chemistry

Applied methods:
Electrochemistry

STEFAN, Raluca-Ioana, Dr.
Department of Analytical Chemistry
Faculty of Chemistry
Blvd. Republicii 13
70346 Bucharest
Phone: +40 1 4102279
Fax: +40 1 4102279
E-Mail: ristef@unichim.sfos.ro

Domains of application:
Analytical chemistry

Applied methods:
Trace analysis of inorganic and organic compounds; separation techniques; electrometric methods; spectrometric methods

---

TANASE, Ion, Dr.
Department of Analytical Chemistry
Faculty of Chemistry
Penduri Str. 90
76229 Bucharest
Phone: +40 1 4102279
Fax: +40 1 4102279

Private address:
Pantelimon Str. 101
Bucharest

Domains of application:
Analytical chemistry in chemical analysis

Applied methods:
Electrometric methods; ion selective electrodes

VLADESCU, Luminita, Dr.
Department of Analytical Chemistry
Faculty of Chemistry
Blvd. Republicii 13
70346 Bucharest
Phone: +40 1 6159249
Fax: +40 1 6159249
E-Mail: vlumi@unichim.sfos.ro

Private address:
Ion Brezoianu Str. 5
70624 Bucharest

Domains of application:
Qualitative and quantitative analysis of inorganic and organic materials; control of the quality of the environmental (water, soil); control of the quality of drugs; synthesis, characterization and utilization of the organic reagents in analytical chemistry

Applied methods:
UV-VIS spectrometric methods; NMR; AAS; potentiometric methods; separation methods based on ion exchange equilibria; chromatography methods

Membership:
International Jury of the International Chemistry Olympiad

Russia

BRAININA, Khiena Zalmanovna, Prof.
Department of Chemistry
Ural State Economic University
8th of March st. 62
620129 Ekaterinburg
Phone: +7 3432 222415
Fax: +7 3432 227147
E-Mail: uinh@uinh.rcupi.e-burg.su

Private address:
A. Valek st. 12
620077 Ekaterinburg

Domains of application:
Environmental monitoring; electroanalytical chemistry

Applied methods:
Stripping voltammetry

Membership:
New York Academy of Sciences; Journal of Solid State Electrochemistry

Awards:
Honored Scientist of Russian Federation

BUDNIKOV, Herman, Dr.
Department of Analytical Chemistry
Kazan State University
Kremlevskaya st. 18
420008 Kazan
Phone: +7 8432 315490
E-Mail: herman.budnikov@ksu.ru

Private address:
Karbyshev st. 40
420087 Kazan

Domains of application:
Searching for new ways to increase selectivity and sensitivity of physical and chemical methods for analysis of small amounts of some biologically active compounds

Applied methods:
Voltammetry; coulometry; amperometry; biosensors; HPLC

BYKOVA, Lidia, Dr.
State Textil Academia
M. Kaluzhskaya 1
119717 Moscow
Phone: +7 95 9553332

Private address:
Dokikina 5

129226 Moscow

Domains of application:
Production of the chemical fibers

Applied methods:
Electrochemical methods

DAVANKOV, Vadim, Dr.
Institute of Organo-Element Compounds
Russian Academy of Sciences
Vavilov str. 28
117813 Moscow
Phone: +7 95 1356471
Fax: +7 95 1356471
E-Mail: davank@ineos.ac.ru

Domains of application:
Liquid chromatography (HPLC); stereochemistry; synthesis and investigation of new adsorbing materials and polymeric adsorbents

Applied methods:
Ligand exchange chromatography; enantiomeric analysis using HPLC; solid phase extraction

Membership:
IUPAC; ACS; Editorial Board of J. Chromatography; Chromatographia; J. Pharm. Biomed. Anal; J. High Resolution Chromatography; J. Biochemical and Biophysical Methods

Awards:
State Awars for Science and Technology, Russi, 1996; Tswett Medal for Achievements in Chromatography, 1978

DAVYDOVA, Svetlana, Dr.
Institute of Petrochemical Synthesis
Russian Academy of Sciences
Leninski Prospect 29
117071 Moscow
Phone: +7 95 9554334
Fax: +7 95 2302224
E-Mail: tips@tips.aha.ru

Private address:
Leninski Prospect 77
117261 Moscow

Domains of application:
Institute is engaged in synthesis and analysis of organic (as well metal-containing organic) compounds for usage as the next: fuels; catalysts of organic reactions; additionals for gazolins; heavy residues of petrols; polymeric membrane separation gases

Applied methods:
AAS for determination of heavy metals in environment; metal and ion selective electrodes for the some aims; determination of ligands (by different methods) leindig trace metals as pollutants

Membership:
International Biographical Centre; International Women Ecological Assembly

---

DOLMANOVA, Inga F., Dr.
Laboratory of Kinetic Methods
Lomonosov Moscow State University
Voroboty Gory
119899 Moscow
Phone: +7 95 9393346
Fax: +7 95 9394675

Private address:
Pudorkin st. 6-3-61
119285 Moscow

Domains of application:
Division of Analytical Chemistry; Department of Chemistry; Lomonosov Moscow State University

Applied methods:
Kinetic methods in analytical chemistry; catalytic and enzymatic methods

---

FILIMONOV, Lev N., Dr.
Giredmet Testing Analytical Center
B. Tomachevsky pers. 5
109017 Moscow
Phone: +7 95 2399211
Fax: +7 95 9561782
E-Mail: analytica@e-mail.girmet.ru

Private address:
Skatertny pers. 21
121069 Moscow

Domains of application:
Theory and applied researches on analytical chemistry of rare metals, semiconductor materials, and high pure substances. Analytical testing and certification of metals, raw materials, recoverable raw, etc.; the environmental objects analysis

Applied methods:
Spectrometry; Spectrophotometry; polarography; titrimetry; gravimetry; gas chromatography; high temperature extraction; activation analysis

---

GILMUTDINOV, Albert, Dr.
Department of Physics
Kazan State University
Kremlevskaya st. 18
420008 Kazan
Phone: +7 8432 315564
Fax: +7 8432 631134
E-Mail: albert.gilmutdinov@ksu.ru

Private address:
Chechova St. 51-70
420043 Kazan

Domains of application:
Chemical spectroscopy

Applied methods:
Atomic absorption spectrometry

Membership:
Member of Editorial Advisory Board of the following Journals: Spectrochimica Acta, Journal of Analytical Atomic Spectrometry; Canadian Journal of Analytical Sciences and Spectroscopy

Awards:
Awarf of Spectrochimica Acta for the most significant contribution published in the Journal in 1994; Soros Associate profesor (1997)

GOLOVNYA, Rimma, Dr. Sci.
Institute of Biochemical Physics
Russian Academy of Sciences
Kosigin str. 4
117977 Moscow
Phone: +7 95 1376420
Fax: +7 95 1374101

Private address:
Lenin Hills Moscow State University
117234 Moscow

Domains of application:
GC of trace organic compound analysis; flavour research

Applied methods:
GC; GC-MS in flavour research

Membership:
European Chemoreception Research Organisation

Awards:
Relationship of structure of organic compounds and sorption for identification

---

GRAZHULENE, Svetlana, Prof.
Institute of Microelectronics Technology and High Purity Materials

Russian Academy of Sciences
142432 Chernogolovka
Phone: +7 95 9622074
Fax: +7 95 9628047
E-Mail: grazhule@ipmt-hpm.ac.ru

Private address:
Central st. 4
142432 Chernogolovka

Domains of application:
Analytical chemistry trace analysis; development of special techniques for the analysis of various objects using the preconcentration technique compatible with modern detection methods; investigation of new sorbents, reagents

Applied methods:
Atomic spectrometry; mass spectrometry; x-ray; neutron activation analysis

---

GRIBOV, Lev, Dr.
Vernadsky Institute of Geochemistry and Analytical Chemistry
Russian Academy of Sciences
Kosygin st. 19
117975 Moscow
Phone: +7 95 1376371
Fax: +7 95 9382054

Private address:
Vernadsky av. 50A
117464 Moscow

Domains of application:
Theory and methods of calculation of molecular spectra; quantum chemistry; molecular modelling; analytical chemistry of compounds; chemical information

Applied methods:
Recognitions of compounds with help of different spectra; detail investigations of ground and excited states of complex moelcules; polymers and crystals; construction of expert systems; IR quantitative analysis of individual organic compounds and mixtures

Membership:
IUPAC

Awards:
Honored Scientist of Russia; Recipient P. Kapitza and Great Peter Honorary Medals

---

IVANOV, Vadim, Dr.
Chemistry Department
Iomonosov Moscow University
Voroboty Gory
119899 Moscow
Phone: +7 95 9392277

Private address:
Komsomolsky Prospect 19-30
119146 Moscow

Domains of application:
Chemistry Department

Applied methods:
Spectrophotometry; organic analytical reagents; HPLC

KAMENEV, Anatoly, Dr.
Department of Chemistry
Moscow State University
119899 Moscow
Phone: +7 95 9391331
Fax: +7 95 9328846

Private address:
Vilnjusskaya str. 8
117574 Moscow

Domains of application:
Electroanalytical chemistry; statistical models; electrochemical devices; chemometrics

Applied methods:
Voltammetry; DC and AC

polarography; coulometry; potentiometry

KARPOV, Yuri A., Dr.
Giredmet Testing Analytical Center
B. Tolmachevsky per. 5
109017 Moscow
Phone: +7 95 2399781
Fax: +7 95 9561782

Private address:
Osenny boulevard 5
121609 Moscow

Domains of application:
Theory and applied researches on analytical chemistry of rare metals, semiconductor materials and high pure substances; analytical testing and certification of metals, raw materials, recoverable raw, etc.; the environmental objects analysis

Applied methods:
Spectrometry; spectrophotometry; polarography; titrimetry; GC; high temperature extraction; activation analysis

Membership:
Accreditation and Quality Assurance Journal for Quality; comparability and reliability in chemical measurements; member of Intercontinental Editorial Board

---

KLYAAHKO, Yurii, Prof.
Institute of Food Industry
Zemlyanoi Val 73
109803 Moscow
Phone: +7 95 9155806

Private address:
Usievicha str. 23
Moscow

Domains of application:
Food industry; biochemistry; food analysis

Applied methods:
Chromatography; electroanalytics; electrophoresis; electromicroscopy

---

KOLESOV, Gennady M., Dr.
Vernadsky Institute of Geochemistry and Analytical Chemistry
Russian Academy of Sciences
Kosygin st. 19
117975 Moscow
Phone: +7 95 9391838
Fax: +7 95 2382054
E-Mail: elkon@geochi.msk.su

Private address:
Lobachevsky str. 18
117415 Moscow

Domains of application:
Methodology of analysis; analysis of environmental object and geochemical and cosmochemical materials

Applied methods:
NAA; AEC (ICP); AAS; BFX; MS

Awards:
Some prizes of Russian Mendelew Society

KOREMAN, Yakov, Dr.
Department of Analytical Chemistry
Voronezh State Academy of Technology
Revolutsii pr. 19
394017 Voronezh
Phone: +8 732 550762
Fax: +8 732 554267
E-Mail: post@vgta.comch.ru

Private address:
Plehanovskaya st. 22
394030 Voronezh

Domains of application:
Elaboration of the theoretical principles of extraction of aromatic

compounds; the methods of multiple-factor prognostication of extraction characteristics on the basis of correlation interrelations; the development of new methods of concentration

Applied methods:
Liquid extraction; gas-liquid chromatography; highly efficient liquid chromatography; spectrophotometry; potentiometric titration

Membership:
The International Academy of Sciences of Ecology and Life Protection (IASELP); The International Editorial Board of the Journal of Serbian Chemical Society; The Editorial Board of the Ecological Congress International Journal

Awards:
The grant of the President of Russian Federation; the title of Soros Professor

---

KUZ'MIN, Nikolai, Prof.
Chemistry Department
Lomonosov Moscow State University
Voroboty Gory
119899 Moscow
Phone: +7 95 9392579
Fax: +7 95 9394675
E-Mail: elhor@geochi.msk.su

Private address:
Teply Stan 14/2
117133 Moscow

Domains of application:
Preconcentration of trace components; intensification of sample preparation; microwave chemistry; environmental analytical chemistry; analysis of high purity materials; analysis of mineral raw material

Applied methods:
Methods of preconcentration; solvent extraction and sorption; atomic emission and atomic absorption; spectrometry; spectrophotometry

---

KUZNETSOV, Vladimir V., Dr. Sci.
Department of Analytical Chemistry
D. Mendeleyer University of Chemical Technology of Russia
Miusskaya Square 9
125047 Moscow
Phone: +7 95 9789196
Fax: +7 95 2004204
E-Mail: kuzn@mhti.msk.su

Private address:
Uralskaya Street 1
107241 Moscow

Domains of application:
Research and teaching on analytical chemistry, advanced courses; PhD Thesis supervision; administration responsabilities; chemical education

Applied methods:
Flow injection analysis; spectrophotometry (optical sensors); chemometrics (graph method, fuzzy logic, neural sets in analytical chemistry)

Awards:
Honorary Chemist of Russian Federation; Soros Professor

LOSER, Nikolai, Dr.
North Caucasus Scientific Centre
Pushkinskaya 140
344006 Rostov-an-Dan
Phone: +7 6382 645597
Fax: +7 6382 640833

Private address:
Kirovskii 44
344020 Rostov-an-Don

Domains of application:
Co-ordination of scientific research activity on the territory of North Caucasus, Caspian Region and South Volga Region, Elaboration of the programs for economic development of North Caucasus and included regions

Applied methods:
X-ray spectroscopy and analysis; the science of the materials

Awards:
The orders and the madaes of the USSR; Honoured man of the Science and Technique of the RSFR

L'VOV, Boris, Dr.
Department of Analytical Chemistry
St. Petersburg State Technical University
Politechnicheskaya ul 29
195251 St. Petersburg
Phone: +7 812 5527741
E-Mail:
lvov@achem.hop.stu.neva.ru

Private address:
Zanevsky Prospekt 43
195213 St. Petersburg

Domains of application:
Atomic absorption spectroscopy; absolute methods of atomic absorption analysis; kinetics and mechanism of solid state reactions

Applied methods:
AAS; electrothermal atomization; thermal analysis

Membership:
IUPAC; Member of the Editorial Advirosy Boards: Spectrochimica Acta Part B (SAP); Journal of Analytical Atomic Spectrometry (JAAS); Canadian Journal Applied Spectroscopy; Spectroscopy and Spectral Analysis

Awards:
Talanta Gold Medal; James Waters Prize; Bunsen-Kirchhoff Prize; Cardinal Leme Medal; Jana Marca Marci Medal; XXX CSI Medal; Doctor of Sciences Honorius Causa Degree

---

MAISTHENKO, Valery, Dr.
Institute for Problem of Applied Ecology and Natural Resources Use
8 Marta Street 12/1
450005 Ufa
Phone: +7 3472 287771
Fax: +7 3472 283913
E-Mail: ipendy@mts.bashkiria.su

Private address:
Stepan Kuvikin Street 1
450103 Ufa

Domains of application:
Analytical chemistry of environmental objects

Applied methods:
Electrochemical methods; chromatography; chromato-mass-spectrometry

---

MOSKVIN, Leonid, Dr.
Department of Analytical Chemistry
Sankt-Petersburg State University
Universitetskii pr. 2
198304 St. Petersburg
Phone: +7 812 4286833
Fax: +7 812 4286939

Domains of application:
Development of separation and concentration methods; spectroscopic; chromatographic; photometric; luminescence; electrochemical continuous flow and

flow injection methods of analysis

Applied methods:
FIA; gas and ion chromatography; atomic absorption; atomic emission and Mössbauer spectroscopy; potentiometry; voltametry; photometry

---

MUZGIN, Vladimir Nikolaavitch, Dr.
Urals State Technical University
Physics-Technical Faculty
Myra Street 19
620002 Ekaterinburg
Phone: +7 3432 448505
Fax: +7 3432 441624
E-Mail: muzgin@pcma.rcupi.e-burg.su

Private address:
Tatischeva Street 77
620131 Ekaterinburg

Domains of application:
Student education and engineer-analyst training; arbitration analysis carrying out and products certification; atomic spectroscopy

Applied methods:
AAS; optical emission spectral analysis; UV-, VIS-, IR-spectroscopy; mass spectroscopy; z-ray spectrometrical analysis; activation and radiochemical analysis

---

MYASOEDOV, Boris F., Dr.
V.I. Vernadsky Institute of Geochemistry and Analytical Chemistry
Russian Academy of Sciences
Kosygin st. 19
117975 Moscow
Phone: +7 95 1374147
Fax: +7 95 9382054
E-Mail: bf@ncrc.msk.ru

Private address:
Ostrovitjanova st. 37
117279 Moscow

Domains of application:
Radiochemistry methods of separation and determination of actinides

Applied methods:
All modern analytical and radiochemical methods

Membership:
IUPAC

NEMETS, Valery, Dr.
Institute of Physics
St. Petersburg State University
Ulyanovskaya 1
198904 St. Petersburg
Phone: +8 812 4287200
Fax: +8 812 4287240
E-Mail:
oshemkov@snoopy.phys.rpbu.ru

Private address:
Suvorovskoya, 15-80
118904 St. Petersburg

Domains of application:
Spectral analysis; emission analysis of gases; isotopic analysis; laser analytical spectroscopy

Applied methods:
Emission spectroscopy; atomic absorption spectroscopy; laser induced fluorescence spectroscopy; laser ablation; electrothermal atomization

---

POPOV, Alexander, Prof.
NPO "Khimavtomatika"
ul. Selskohozyaistvennaya
129226 Moscow
Phone: +7 95 1811684
Fax: +7 95 1816012

Private address:
Tzehprudni per 11/13
103001 Moscow

Domains of application:
Analytical measurements

Applied methods:
Gas and ion chromatography; x-ray fluorescents and other methods

Awards:
The Prize of Russian Government

RAMENDIK, Gregory, Dr. Sci.
Kurnasov Institute of General and Inorganic Chemistry
Russian Academy of Sciences
Leninsky Prospect 31
117907 Moscow
Phone: +7 95 2364277
Fax: +7 95 9541279
E-Mail: ramendik@ionchran.msk.ru

Private address:
Vavilow st. 6
117312 Moscow

Domains of application:
My laboratory is developing analytical methods and fulfilling chemical analysis of organic substances, produced or used in the Institute

Applied methods:
Mass-spectrometry with plasma ion sources; GC-MS; chromatography; x-ray analysis; atomic absorption; CHNOS-analysis

Membership:
IUPAC; EURACHEM; Cooperation on International Traceability in Analytical Chemistry

---

REVELSKY, Igor, Dr. Sci.
Department of Chemistry
Moscow State University
Voroboty Gory
119899 Moscow
Phone: +7 95 9394318
Fax: +7 95 9394675
E-Mail:
revelsky@environment.chem.msu.su

Private address:
Trehprudny Per. 8/6
Moscow

Domains of application:
GC; HPLC; IC; HPTLC; GC/MS; laser micro-mass spectrometry; SFE; environmental pollution control; chemical products and materials quality control; based on combine use of these methods

---

RUDENKO, Boris, Dr.
Vernadsky Institute of Geochemistry and Analytical Chemistry
Russian Academy of Sciences
Kosygin st. 19
117975 Moscow
Phone: +7 95 1377625
Fax: +7 95 9372054

Private address:
Paustovsky Str. 4
Moscow

Domains of application:
Gas, liquid; ion chromatography; theory and applications of ion exchange; water desalination and mineral components recovery

Applied methods:
Gas, liquid, ion chromatography; atomic absorption; radiactive indicator technique

Awards:
Honourable M. S. Tswett Medal of Scientific Council on Chromatography, 1978

RUNOV, Valentin, Prof.
Department of Chemistry, Analytical Chemistry Division
Lomonosov Moscow State University
Voroboty Gory
119899 Moscow
Phone: +7 95 9394648
Fax: +7 95 9394675

Private address:
Ostozhenta 30
119034 Moscow

Domains of application:
Fundamental analytical chemistry: molecular, atomic, x-ray spectroscopy, thermo-optical spectroscopy, HPLC, ion chromatography, kinetic and catalytic methods of analysis, extraction and soprtion preconcentration methods; polarography; voltametry

Applied methods:
Molecular, atomic, x-ray, x-ray electron and thermo-optical spectroscopy; applied analysis of environmental materials and platinum-group metals

---

SALITHDZANOVA, Rashida, Dr.
Moscow Institute of the Radio Engineering, Electronic and Automation
pr. Vernadscogo 78
117454 Moscow
Phone: +7 95 4349229
Fax: +7 95 4348665

Private address:
Paveleckay 1
113114 Moscow

Domains of application:
Technology and production of radio and electronics

Applied methods:
Voltammetry; voltammetric equipments

SAVVIN, Sergey, Dr.
Institute of Geochemistry and Analytical Chemistry
Russian Academy of Sciences
Kosygin st. 19
117975 Moscow
Phone: +7 95 1372878
Fax: +7 95 9382054

Domains of application:
Organic reagents in analytical chemistry

Applied methods:
Spectrophotometry

SHEKHOVTSOVA, Tatyana, Dr. Sci.
Chemistry Department
Lomonosov Moscow State University
Voroboty Gory
119899 Moscow
Phone: +7 95 9393346
Fax: +7 95 9394675
E-Mail:
shekhov@chromat.chem.msu.su

Private address:
Mosfilmovskaya st. 15
117330 Moscow

Domains of application:
Kinetic and enzymatic methods in chemical analysis; fundamentals and application: environmental analysis, medicine, biochemical research

Applied methods:
Spectroscopy; chromatography

SHPIGUN, Lihja, Dr.
Kurnakov Institute of General and Inorganic Chemistry

Russian Academy of Sciences
Leninsky Prospect 31
117907 Moscow
Phone: +7 95 9521429
Fax: +7 95 9541279

Private address:
Ostrovityanov st. 29
117234 Moscow

Domains of application:
Analytical chemistry

Applied methods:
Electroanalytical methods; catalytic methods of analysis; membrane sensors; spectrophotometry; flow injection analysis

Membership:
IUPAC

SHPIGUN, Oleg, Prof.
Chemistry Department
Lomonosov Moscow State University
Voroboty Gory
119899 Moscow
Phone: +7 95 9321382
Fax: +7 95 9394675
E-Mail:
shpigun@chromat.chem.msu.su

Private address:
Mosfilmovskaya 15-44
117330 Moscow
Domains of application:
Determination of anions and cations in environmental samples

Applied methods:
Ion chromatography; CE

Awards:
Russian State Prize in Science

SHTYKOV, Sergei, Dr. Sci.
Department of Chemistry
Saratov State University
Astrakhanskaya 83
410026 Saratov
Phone: +7 8452 245436
Fax: +7 8452 240446
E-Mail: shtykov@scnit.saratov.su

Private address:
Blagodorova Street 9
410086 Saratov

Domains of application:
Analytical chemistry and chemical ecology; physical chemistry (electrochemistry); organic chemistry and biotechnology; inorganic chemistry; physicochemistry of polymers; technical chemistry

Applied methods:
Absorption and emission molecular; NMR-spectroscopy; thin layer (TLC); HPLC; physichemical properties and application of microheterogeneous organized media (micelles, microemulsions, cyclodextrins) in analytical chemistry

Membership:
IUPAC

Awards:
1994-96 Holder of Stete Grant for Outstanding Scientist of Russia and 1997-1999 also

---

TERENT'EVA, Evgenia, Dr.
Nesmeyanov Institut of Organoelement Compounds
Russian Academy of Sciences
Vavilov str. 28
117813 Moscow
Phone: +7 95 1359230
Fax: +7 95 1355085
E-Mail: manlab@ineos.ac.ru

Private address:
Prospekt Mira 71
129110 Moscow

Domains of application:
Chemistry of organoelement compounds and polymers

Applied methods:
Ion chromatography; polarography; coulometry; spectrophotometry; AAS; visual titrimetry; express-gravimetry; RX-fluorescence spectrometry

Membership:
IUPAC;

Awards:
Russian Mendelsev Chem. Soc., 1974; Allunion Exhibition of National Economy USSR (silver medal); Inst. of Organoelement Comp. (2 annual awards)

TRUBACHEV, Aleksey, Dr. Sci.
Institute of Applied Mechanics of Ural Branch RAS
Georkii str. 222
426001 Izhevsk
Phone: +7 3412 250188
Fax: +7 3412 231713
E-Mail: ipm@ipm.udm.ru

Private address:
Krasnoarmeyskaha str.
426057 Izhevsk

Domains of application:
Mechanics of liquids and gases; physical material science; instrumental problems on the interior ballistic systems integrated by solid propellants; mathematic simulation of processes; analysis of materials and new methods and apparatus for control

Applied methods:
Parallel system calculation methods; x-ray fluorescence methods; mass spectrometry methods; electron-spectroscopy methods; chromatographic methods; voltammetric methods; spectrophotometric methods

Membership:
IUPAC; International Association for Fire Safety Science

TSYSIN, Grigory, Dr.
Chemistry Department
Moscow State University
Voroboty Gory
113799 Moscow
Phone: +7 95 9395518
E-Mail: root@conc.chem.msu.su

Private address:
Bol. Rogogsky per. 10
Moscow

Domains of application:
Analytical chemistry

Applied methods:
Sorption preconcentration of metals; synthesis of sorbents; flow-injection systems; FAAS; FAF; XRF; AES-ICP determination of elements

---

ULAKHOVICH, Nikolay, Dr.
Department of Inorganic Chemistry
Kazan State University
Kremlevskaya st. 18
420008 Kazan
Phone: +7 8432 315416
Fax: +7 8432 380994
E-Mail: nikolay.ulakhovich@ksu.ru

Private address:
Zarya 24-23
420029 Kazan

Domains of application:
Structure and reactivity of organic, organometal and coordination compounds

Applied methods:
Voltammetry; spectrometry; potentiometry; IR; NMR;

spectroscopy

Awards:
Professor of Soros

VARSHAL, Galina M., Dr.
V.I. Vernadsky Institute of Geochemistry and Analytical Chemistry
Russian Academy of Sciences
Kosygin st. 19
117975 Moscow
Phone: +7 95 1377526
Fax: +7 95 9382054

Private address:
Vavilova str. 46
117333 Moscow

Domains of application:
Analytical chemistry and geochemistry of environment; chemical speciation of elements in natural waters, soils, river and marine sediments; the mechanism of interaction of metal ions with humus acids and carbonaceous matter of rocks

Applied methods:
Atomic absorption spectrometry; atomic emission spectrometry with ICP; cold vapor atomic fluorescence spectrometry; sorption-AAS methods; sotpion-AES-ICP methods; sorption-XRF-spectrometry methods; kinetic methods; CHNS analysis

Awards:
Award of Cabinet Council of Russian Federation, 1987

VLASOV, Yuri Georgievich, Prof.
Departmento and Laboratory of Chemical Sensors
St. Petersburg State University
Universitetskaya nab. 7/9
199034 St. Petersburg
Phone: +7 812 2139731
Fax: +7 832 2182835
E-Mail:
andrew@sensor.chem.lgu.spb.su

Domains of application:
Inorganic chemistry; physical chemistry; organic chemistry; radiochemistry; analytical chemistry; electrochemistry; natural compound chemistry; solid state chemistry

Applied methods:
Potentiometry; Mössbauer spectroscopy; conductivity; radioactive isotope; pattern recognition; electronic Tongue

Membership:
IUPAC; Memer of International Stirring Committee of Eurosensor

Awards:
Kapiza Award of the Royal Society, 1992; Soros Professor, 1994

VLASOV, Yuri Georgiewich, Prof.
St. Petersburg State University
Universitetskaya nab. 7/9
199034 St. Petersburg
Phone: +7 812 2189731
Fax: +7 831 2182835
E-Mail:
andrew@sensor.chem.lgu.spb.su

Domains of application:
Inorganic chemistry; physical chemistry; organic chemistry; radiochemistry; analytical chemistry; electrochemistry; natural compound chemistry; solid state chemistry

Applied methods:
Potentiometry; Mössbauer spectroscopy; conductivity; radioactive isotope; patterns recognition; electronic tongue

Membership:
IUPAC; EUROSENSOR

Awards:
Kapiza Award of the Royal Society, 1992; Soros Professor, 1994

ZOLOTOV, Yuri, Dr.
Kurnasov Institute of General and Inorganic Chemistry
Russian Academy of Sciences
Leninsky Prospect 31
117907 Moscow
Phone: +7 95 9395564
Fax: +7 95 9541279
E-Mail: zolotov@ionchr.dol.ru
Private address:
Leninskii prospekt 69
117296 Moscow

Domains of application:
Environmental analysis (waters); analysis of high putity materials; analysis of semiconductors; analysis of technological solutions

Applied methods:
Flow injection analysis; sorption; liquid-liquid extraction; UV-vis spectrophotometry; ion chromatography; HPLC; AAS; express test methods

Membership:
DAC-FECS; International Committee on Solvent Extraction Chemistry and Technology

Awards:
F. Emich Award (1999; Honorary Foreign Member of Japan Soc. for Anal. Chem. (1991); L. Gordon Award (1988); USSR State Prize (1972); Russia Federation State Prize (1991); Mendeleev Gold Medal (1993)

Slovak Republic

BEREK, Dusan, Dr. Sci.
Polymer Institute of the Slovak Academy of Sciences
Dubravska cesta 9
84236 Bratislava
Phone: +421 7 3782306
Fax: +421 7 375923
E-Mail: upoldber@sauba.sk

Private address:
Medena 27
81102 Bratislava

Domains of application:
Basic research on polymer modification, degradation and characterization; molecular characterization of complex polymers by coupled liquid chromatographic techniques; HPLC/SEC column packings; SPE carbonaceons sorbents

Applied methods:
SEC (GPC); HPLC; SPE

Membership:
IUPAC; FECS; COST

BRANDSTETEROVA, Eva, Assoc. Prof., Dr.
Department of Analytical Chemistry
Faculty of Chemical Technology
Radlinskeho 9
81237 Bratislava
Phone: +421 7 326043
Fax: +421 7 393198
E-Mail: branstet@cvt.stuba.sk

Private address:
Robotnicka 10
82103 Bratislava

Applied methods:
HPLC; therepeutic drug monitoring, pharmacology, clinical analysis

Membership:
International Union against Cancer

BUSTIN, Dusan, Dr. Sci.
Department of Analytical Chemistry
Faculty of Chemical Technology
Radlinskeho 9
81237 Bratislava
Phone: +421 7 395217
Fax: +421 7 393198
E-Mail: bustin@cvt.stuba.sk

Private address:
Nezabudkova 8
82101 Bratislava

Domains of application:
Analytical chemistry

Applied methods:
Electrochemical; spectroscopic; chromatographic

Membership:
ACS; IUPAC

GARAJ, Jan, Dipl. Eng., Dr. Sci.
Department of Analytical Chemistry
Slovak Technical University
Radlinskeho 9
81237 Bratislava
Phone: +421 7 326043
Fax: +421 7 393198

Private address:
Stare Grunty-Sitiny 194
84100 Bratislava

Domains of application:
Electroanalytical chemistry; separation technics (GC, HPLC, isotachophoresis); x-ray diffraction; spectroanalytical chemistry (AAS; ICP); ultrapure laboratory for trace analysis

Membership:
EURACHEM; DAC-FECS

Awards:
Gold Medal of Chemical Faculty;
Gold Medal of Technical University

HUTTA, Milan, Dr., Assoc. Prof.
Department of Analytical Chemistry
Faculty of Natural Sciences
Mlynska Dolina 2
84215 Bratislava
Phone: +421 7 796506
Fax: +421 7 725360
E-Mail: hutta@fns.uniba.sk

Domains of application:
LC (HPLC, micellar LC) techniques; trace analysis; metal speciation; environmental analysis; complex matrices (environmental, food, clinical)

Applied methods:
Electromigration, liquid chromatography methods; atomic spectrometry methods
Membership:
International Humic Substances Society

KANTANSKY, Dusan, Dr., Assoc. Prof.
Department of Analytical Chemistry
Faculty of Natural Sciences
Mlynska Dolina 2
84215 Bratislava
Phone: +421 7 796506
Fax: +421 7 725360
E-Mail: kaniansky@fns.uniba.sk

Domains of application:
Capillary electroseparation techniques; trace analysis; environmental analysis; bioanalysis

Applied methods:
Various variants of CE; LC; GC-MS

KOSTURIAK, Adam, Dr.
Department of Physical and

Analytical Chemistry
Faculty of Sciences
Moyzesova 11
04154 Kosice
Phone: +421 95 6222605

Private address:
Urbankova 31, 04100 Kosice

Domains of application:
The application of new organic agents on basis of 2,3-dioxoindole for proof and determination of metal ions; the application of high temperature organic magnetics in: electrotechnic transformation and transfer of electric energy; agriculture

Applied methods:
UV-VIS; IR; x-ray spectrophotometry; potentiometry; polarography; chemometrics ; Mössbauer spectrometry; magnetising curves; magnetic permeability

---

KOZAKOVA, Eva, Dr., Assoc. Prof.
Department of Analytical Chemistry
Faculty of Natural Sciences
Mlynska Dolina 2
84215 Bratislava
Phone: +421 7 796344
Fax: +421 7 725360
E-Mail: kozakova@fns.uniba.sk

Private address:
Svidnicka 9, 82103 Bratislava

Domains of application:
Equilibrium in non aqueous solvents; measurements of pH in mixed and non aqueous solvents; environmental analysis; speciation of heavy metal in soils

Applied methods:
Electroanalytical chemistry; stripping analysis; potentiometric analysis

---

LABUDA, Jan, Dipl. Eng., Dr.
Department of Analytical Chemistry
Slovak Technical University
Radlinskeho 9
81237 Bratislava
Phone: +421 7 326021
Fax: +421 7 393198
E-Mail: labuda@chtf.stuba.sk

Domains of application:
Electrochemical methods of analysis; metal speciation analysis

Applied methods:
Voltammetric analysis (DPP, ASV, CV); biosensors and chemically modified electrodes; flow-through systems

---

MADAJOVA, Vlasta, Dr., Assoc. Prof.
Department of Analytical Chemistry
Faculty of Natural Sciences
Mlynska Dolina 2
84215 Bratislava
Phone: +421 7 796379
Fax: +421 7 725360
E-Mail: analytika@fns.uniba.sk
Private address:
Astrova 16
82101 Bratislava

Domains of application:
Trace analysis of multicomponent mixtures; environmental analysis

Applied methods:
Electrophoretic separation methods; isotachophoresis (ITP); capillary zone electrophoresis (CRE); ITP-CZE

MASAR, Marian, M. Sci.
Department of Analytical Chemistry
Faculty of Natural Sciences

Mlynska Dolina 2
84215 Bratislava
Phone: +421 7 796379
Fax: +421 7 725360
E-Mail: masar@fns.uniba.sk

Private address:
Podolie 329
91622 Podolie

Domains of application:
Environmental analyses; food products

Applied methods:
Capillary electrophoresis: isotachophoresis (ITF); capillary zone electrophoresis (ZE); on-line combination ITF-CZE

SIMUNICOVA, Eva, Dr.
Department of Analytical Chemistry
Faculty of Natural Sciences
Mlynska Dolina 2
84215 Bratislava
Phone: +421 7 796379
Fax: +421 7 725360
E-Mail: analytika@fns.uniba.sk

Private address:
Gabrikova 2
84105 Bratislava

Domains of application:
Trace analysis of multicomponents; mixtures (anorganic, organic and biological materials); environmental analyses

Applied methods:
Electrophoresis separation methods; isotachophoresis (ITP); capillary zone electrophoresis (CZE); ITP-CZE

STEFUCA, Vladimir, Dipl. Eng., Dr.
Department of Chemical and Biochemical Engineering
Faculty of Chemical Technology
Radlinskeho 9
81237 Bratislava
Phone: +421 7 396920
Fax: +421 7 396743
E-Mail: stefuca@cvt.stuba.sk

Private address:
Hany Melickovej 20
84105 Bratislava

Domains of application:
Biotransformation; fermentations

Applied methods:
Enzyme flow microcalorimetry; HPLC

STELTENPOHL, Pavol, Dr.
Department of Chemical and Biochemical Engineering
Faculty of Chemical Technology
Radlinskeho 9
81237 Bratislava
Phone: +421 7 326021
E-Mail: steltemp@cvt.stuba.sk

Private address:
Bakosova 44
84102 Bratislava

Domains of application:
Chemical and biochemical engineering; trace analysis; gas chromatography; characterization of porous materials

Applied methods:
GC; GC-MS; UV-VIS spectroscopy, mercury porosimetry; adsorption of nitrogen on porous materials

---

ZELENSKA, Viera, Dr.
Department of Analytical Chemistry
Faculty of Natural Sciences
Mlynska Dolina 2
84215 Bratislava
Phone: +421 7 796379

Private address:
Pupanona 5
84104 Bratislava

Domains of application:
Trace analyses of multicomponent mixtures

Applied methods:
Electrophoretic separation methods

ZEMBERYOVA, Maria, Dr.
Department of Analytical Chemistry
Faculty of Natural Sciences
Mlynska Dolina 2
84215 Bratislava
Phone: +421 7 796309
Fax: +421 7 725360
E-Mail: zemberyova@fns.uniba.sk

Private address:
Mudrochova 3
83106 Bratislava

Domains of application:
Speciation of heavy metals in soils and soil extracts; determination of mercury and methylmercury; determination of toxic metals in biological matrices

Applied methods:
Atomic absorption spectrometry

Spain

ALAÑÓN MOLINA, Aurelia, Dr.
Department of Analytical Chemistry and Food Technology
Faculty of Sciences
Av. Camilo José Cela, 10
13071 Ciudad Real
Phone: +34 926 295300
Fax: +34 926 295318
E-Mail: aalanon@qata-cr.uclm.es

Domains of application:
Drugs analysis; environmental analysis

Applied methods:
Drugs analysis; environmental analysis

ALBERO QUINTO, María Isabel, Dr.
Department of Analytical Chemistry
University of Murcia
Campus de Espinardo
30071 Murcia
Phone: +34 968 307100
Fax: +34 968 364148
E-Mail: mialbero@fcu.um.es

Private address:
José Castaño, 8
30002 Murcia

Domains of application:
Ion selective electrodes; optical sensors; piezoelectric sensors; FIA; ion transfer across the interface between two immiscible electrolyte solutions by cyclic voltammetry; spectrophotometric and spectrofluorimetric methdos

---

ALEGRET, Salvador, Dr.
Department of Chemistry
Universitat Autónoma de Barcelona
08193 Bellaterra
Phone: +34 93 5811976
Fax: +34 93 5812477
E-Mail: iqan0@cc.uab.es

Domains of application:
Chemical sensors; biosensors; process analytical monitoring; analytical instrumentation

Applied methods:
Immobilization (chemical and biological); biocomposites; modified electrodes; modified magnetic

particles; electrochemical transducers; solid-state transducers; integrated optical transducers; flow injection analysis; thick-film sensor technologies

---

ALLER, A. Javier, Dr.
University of León
Campus de Vegazana, s/n
24071 León
Phone: +34 987 291536
Fax: +34 987 291536
E-Mail: dbbjaf@unileon.es

Domains of application:
Separation of heavy metals using biological systems; atomization mechanisms in a grafite tube

Applied methods:
ETAAS; ETAES; FAAS; FTIR

Membership:
Society for Applied Spectroscopy

Awards:
Special Mention of the Prof. Antonio Hidalgo Award, Perkin Elmer, 1995

ALONSO, Rosa María, Dr.
Departamento de Química Analítica
Universidad del País Vasco
Facultad de Ciencias
48080 Bilbao
Phone: +34 94 4647700
Fax: +34 94 4648500
E-Mail: qapalror@lg.ehu.es
Private address:
Avda. Los Chopos, 43
48990 Getxo

Domains of application:
Recovery of metals; analytical chemistry; doping control; automatization; food analysis; pharmaceutical analysis; environmental chemistry

Applied methods:
Atomic spectroscopic methods; chromatographic methods; capillary electrophoresis; electroanalytical methods

Membership:
RSC

ANDRADE GARDA, José Manuel, Dr.
Department of Analytical Chemistry
University of A Coruña
Campus de Zapateira, s/n
15071 A Coruña
Phone: +34 981 167000
Fax: +34 981 167065
E-Mail: Andrade@udc.es

Domains of application:
Environmental studies; applied spectroscopy; applied chromatography; applied chemometrics

Applied methods:
UV-VIS; FT-MIR; FAAS; GFAAS; chromatography (GC, HPLC)

Awards:
2nd International Award of Young Chemometris; 1st Prize of the Portuguese-Galician Award of Young Chemists

---

ANTICÓ, M. Enriqueta, Dr.
University of Girona
Campus Montilivi
17071 Girona
Phone: +34 972 418276
Fax: +34 972 418150
E-Mail: dgead@fc.udg.es

Domains of application:
Separation processes: solvent extraction, liquid membranes, resins; electrodes and optodes; environmental analysis

Applied methods:
AAS; AAS-GF; GC; UV-visible spectroscopy

ANZANO, Jesús, Dr.
Department of Analytical Chemistry
University of Zaragoza
Miguel Servet, 177
50013 Zaragoza
Phone: +34 976 761634
Fax: +34 976 761612
E-Mail: janzano@posta.unizar.es

Domains of application:
Electrotherme atomization atomic absorption spectroscopy; laser induced fluorescence; direct analysis of metals in solid samples by laser excited atomic fluorescence spectrometry using graphite furnace; direct analysis of metla sin food by GFAAS

Applied methods:
Determination of metals in PVC by atomic absorption spectrometry; analysis of foods (migration study)

Membership:
SEQA

ARIÑO, Cristina, Dr.
University of Barcelona
Diagonal, 647
08028 Barcelona
Phone: +34 93 4022545
Fax: +34 93 4022233
E-Mail: cristina@zeus.qui.ub.es

Private address:
Trull, 19
08754 El Papiol

Domains of application:
Environmental chemistry; control analysis; fundamentals of analytical chemistry; chemometrics

Applied methods:
Modern polarographic techniques; stripping analysis; application of chemometric techniques to electroanalysis

Membership:
SEQA; Real Sociedad Española de Química

ARRANZ GARCÍA, Adela, Dr.
Department of Analytical Chemistry
University of País Vasco
Paseo de la Universidad, 7
01006 Vitoria
Phone: +34 945 183057
Fax: +34 945 130756
E-Mail: qapargaa@vf.ehu.es

Domains of application:
Environment; electroanalysis of pesticides and pharmaceuticals; capillary electrophoresis of pharmaceuticals and pesticides

Applied methods:
Voltammetry (polarography); capillary electrophoresis; atomic absorption spectrometry; FIA, UV-vis; potentiometry

---

ARRANZ VALENTÍN, Juan Francisco, Dr.
Department of Analytical Chemistry
University of País Vasco
Paseo de la Universidad, 7
01006 Vitoria
Phone: +34 945 183058
Fax: +34 945 130756
E-Mail: qaparvaj@vf.ehu.es

Domains of application:
Environment; electroanalysis of pesticides and pharmaceuticals; capillary electrophoresis of pesticides and pharmaceuticals

Applied methods:
Voltammetry (polarography); CE;

Spain

AAS; UV-vis; potentiometry; FIA

AYALA MONTORO, Luis, Dr.
Departamento de Física Aplicada
Universidad de Málaga
Campus de Teatinos, s/n
29071 Málaga
Phone: +34 95 2131928
Fax: +34 95 2131124
E-Mail: layala@coie.uma.es

Private address:
Blanco Coris, 6
29007 Málaga

Domains of application:
Espectroscopía láser

Applied methods:
Ablación por láser; fosforimetría inducida láser

BAEZA BAEZA, Juan José, Dr.
Department of Analytical Chemistry
University of Valencia
Dr. Moliner, 50
46100 Burjasot
Phone: +34 96 3983176
Fax: +34 96 3864436
E-Mail: juan.baeza@uv.es

Domains of application:
Chemometrics; chromatography-HPLC; spectroscopy; kinetics

Membership:
SEQA

Awards:
Primer Premio Kurt Kavf, 1989; Chemistry-PhD Award, Univ. Valencia, 1991; Accessit SAQ Award Young Scientists, 1993

BARRADO, Enrique, Dr.
Departamento de Química Analítica
Universidad de Valladolid
Prado de la Magdalena, s/n
47005 Valladolid
Phone: +34 983 423595
Fax: +34 983 423013
E-Mail:
ebarrado@wamba.cpd.uva.es

Private address:
Sargento Provisional, 7
47013 Valladolid

Awards:
Environmental-Chemometric-Food

BARRIO, Ramón J., Dr.
Department of Analytical Chemistry
University of the Basque Country
Paseo de la Universidad, 7
01006 Vitoria
Phone: +34 945 183055
Fax: +34 945 134762
E-Mail: qapbadir@vc.ehu.es

Domains of application:
Research and education

Applied methods:
Voltammetry; microelectrodes; GD-MS; LC-DAD; LC-ED

BARROSO, C.G., Prof., Dr.
Department of Analytical Chemistry
University of Cádiz
Campus Universitario Río San Pedro
11510 Puerto Real
Phone: +34 956 830251
Fax: +34 956 834924
E-Mail: carmelo.garcia@uca.es

Domains of application:
Wine analysis; vinegard analysis; brandy analysis; phenolic compounds; organic acids; aroma; browning; chemometric; automation

Applied methods:
HPLC; MS; GC: HPLC-MS; EC; GC-MS; GC-LV; sample preparation

Membership:
Group of Polyphenols

BASADRE PAMPÍN, María Isabel, Dr.
Department of Analytical Chemistry
Universidad de Santiago
Avda. de las Ciencias, s/n
15706 Santiago de Compostela
Phone: +34 981 563100
Fax: +34 981 547141
E-Mail: qnmaria@usc.es

Private address:
Santa Marta, 4-6
15706 Santiago de Compostela

Domains of application:
Analytical chemistry

Applied methods:
Atomic spectroscopy; molecular spectroscopy; chromatography

BELTRÁN LUCENA, Rafael, Dr.
Departamento de Química y Ciencia de los Materiales
Universidad de Huelva
Campus de La Rábida
21819 Palos de la Frontera
Phone: +34 959 530213
Fax: +34 959 350311
E-Mail: beltran@uhu.es

Domains of application:
Speciation of organometallic compounds in environmental samples (Sn, As, Se, Hg); mobility of heavy metals from sediments (sequential extractions schemes); environmental reference materials

Applied methods:
Atomic spectrometry (AAS, GFAAS, ICP-MS); chromatography (GC, HPLC); coupled methods (GC-FPD, GC-MS, HPLC-UV-HG-AAS)

BERMEJO BARRERA, Adela, Dr.
Department of Analytical Chemistry
University of Santiago
Avda. de las Ciencias, s/n
15706 Santiago de Compostela
Phone: +34 981 563100
Fax: +34 981 595012
E-Mail: qnadbaba@usc.es

Domains of application:
Analytical chemistry

Applied methods:
Water analysis; environmental samples; biological samples

BERRUETA SIMAL, Luis Ángel, Dr.
Departamento de Química Analítica
Facultad de Ciencias
Apartado 644
48080 Bilbao
Phone: +34 944 647700
Fax: +34 944 648500
E-Mail: qapbesil@lg.ehu.es

Domains of application:
Biopharmaceutical analysis; food analysis; environmental analysis

Applied methods:
Chromatography; spectroscopy; electrochemistry

BLANCO ROMÍA, Marcelo, Prof.
Departamento de Química
Universitat Autónoma de Barcelona
Facultad de Ciencias
08193 Bellaterra
Phone: +34 93 5811367
Fax: +34 93 5812477
E-Mail: iqan@blues.uab.es

Private address:
Avda. Josep Tarradellas, 17
08029 Barcelona

Spain

Domains of application:
Molecular spectroscopy; near infrared spectroscopy (NIR); UV-vis; fluorescence; IR-FT; capillary electrophoresis; kinetic analysis; chemometrics

Applied methods:
Pharmaceutical analysis; textiles; petrochemical products

Membership:
Society for Applied Spectroscopy; SEQA; Real Sociedad Española de Química

---

BOSCH, Elisabeth, Dr.
University of Barcelona
Diagonal, 647
08028 Barcelona
Phone: +34 93 4021284
Fax: +34 93 4021233
E-Mail: bosch@zeus.qui.ub.es

Domains of application:
Solution chemistry; chemometrics; chromatography and related techniques

Membership:
IUPAC; ACS; Correlation Analysis in Chemistry

BOSQUE SENDRA, Juan Manuel, Dr.
Department of Analytical Chemistry
University of Granada
Fuetenueva, s/n
18071 Granada
Phone: +34 958 243399
Fax: +34 958 243328
E-Mail: jbosque@goliat.ugr.es

Domains of application:
Quality assurance; chemometrics; toxicological food analysis; environmental analysis; pharmaceutical and biomedical analysis

Applied methods:
Spectrophotometry; luminescence; chromatography techniques

BOTANA LÓPEZ, Ana María, Dr.
Department of Analytical Chemistry
University of Santiago
Augas Ferreas, s/n
27002 Lugo
Phone: +34 982 223996
Fax: +34 982 224904
E-Mail:
botanaqa|secuslugo.lugo.usc.es

Domains of application:
Food and environmental technology

Applied methods:
HPLC; capillary electrophoresis; field-flow fractionation; gas chromatography

---

CACHO, Juan, Dr.
Department of Analytical Chemistry
Faculty of Sciences
Ciudad Universitaria
50009 Zaragoza
Phone: +34 976 761177
Fax: +34 976 761292
E-Mail: jcacho@posta.unizar.es

Private address:
Paseo de las Damas, 5
50008 Zaragoza

Domains of application:
Teaching analytical chemistry; research and development in the field of wine analytical chemistry as well as packaging polymeric materials in contact with food

Applied methods:
GC-FID; ECD; NIP; S-D; MS/MS; olfactometric detector; HPLC-UV; fluorescence; RI; MS

Membership:
RSC; Sociedad Española de Química

Analítica

Awards:
1st Price to the Best Research Work in the 10th International Enomag. Zaragoza, Spain, 1993; 1st Price to the Best Research Work in the Symposium on Analytical Methodology in the Environmental Field, Bilbao, Spain, 1996

CALVO BLÁZQUEZ, Lorenzo, Dr.
Departmento de Química Analítica y Electroquímica
Universidad de Extremadura
Avda. de Elvás, s/n
06071 Badajoz
Phone: +34 924 289392
Fax: +34 924 274244
E-Mail: lorcalvo@ba.unex.es

Domains of application:
Trace organics in environmental and biological samples; trace elements in environmental and biological samples; liquid chromatography with electrochemical detection; voltammetry; atomic absorption; gas chromatography with MS detection

Applied methods:
Voltammetric methods for drugs and trace elements in real samples

---

CÁMARA, Carmen, Dr.
Department of Analytical Chemistry
University Complutense
Ciudad Universitaria
28040 Madrid
Phone: +34 91 3944338
Fax: +34 91 3944329
E-Mail:
ccamara@euchax.sim.ucm.es

Domains of application:
Environment; food

Applied methods:
Atomic spectroscopy: AAS, MIP, ICP-MS, AFDES; capilar electrophoresis; HPLC; GC

Membership:
RSC

CANALS, Antonio, Dr.
Departamento de Química Analítica
Universidad de Alicante
Facultad de Ciencias
03080 Alicante
Phone: +34 96 5903529
Fax: +34 96 5903464
E-Mail: antonio.canals@ua.es

Domains of application:
MW: instrumental design, MW-assisted chemistry; fundamental studies on nebulization and desolvation in atomic spectrometry; physical and chemical characterization of aerosols: particle size distribution characterization and metal determination

Applied methods:
AAS; electrothermal atomization, hydride generation; cold vapor; inductively coupled plasma atomic emission spectrometry; particle size characterization; infrared desolvation; microwave digestion and extraction for metal

Membership:
SEQA; Grupo Espectroquímico de la Real Sociedad Española de Química; Society for Applied Spectroscopy

---

CANO PAVÓN, José Manuel, Dr.
Department of Analytical Chemistry
University of Málaga
Campus de Teatinos
29071 Málaga
Phone: +34 952 131887
Fax: +34 952 132000

Spain

Domains of application:
ICP emission spectrometry; atomic absorption spectrometry; UV-vis spectrophotometry; FIA; liquid chromatography; chemometrics; molecular fluorescence spectrometry; environmental analysis; history of the sciences (social, institutional and scientific)

Applied methods:
Analysis and speciation of metal ions, pesticides, waters, environmental samples and biological samples

Membership:
Real Sociedad Española de Química; SEQA; RSC

CAPITÁN VALLVEY, Luis Fermín, Dr.
Department of Analytical Chemistry
University of Granada
Avda. Fuentenueva, s/n
18071 Granada
Phone: +34 958 243388
Fax: +34 958 243328
E-Mail: lcapitan@goliat.ugr.es

Domains of application:
Solid phase spectrophotometry; spectrofluorimetry; phosphorimetry; HPLC; GC-MS

Applied methods:
Environmental analysis; food and cosmetic analysis; cultural heritage analysis

---

CÁRDENAS ARANZANA, M. Soledad, Dr.
Department of Analytical Chemistry
University of Córdoba
Edificio C-3 Anexo. Campus de Rabanales
14071 Córdoba
Phone: +34 957 211066
Fax: +34 957 218616
E-Mail: qa1caarm@uco.es

Domains of application:
Automatic samples preparation; flow injection analysis; chromatography; screening systems

Applied methods:
Automatic methods in the field of toxicology and pharmacology; adulterants and contaminants in foods

Awards:
Price of Excellence in PhD

CARO GÁMEZ, Concepción, Dr.
Departamento de Didáctica de las Ciencias Experimentales y de las Matemáticas
Universidad de Extremadura
Avda. de Elvás, s/n
06071 Badajoz
Phone: +34 924 289486
Fax: +34 924 274463

CASTAÑEDA, Gregorio, Dr.
Department of Analytical Chemistry
Faculty of Sciences
Campus Universitario, s/n
13003 Ciudad Real
Phone: +34 926 295300
Fax: +34 926 295318
E-Mail: qata-cr@correo.qata-cr.uclm.es

Domains of application:
Drugs analysis; environmental analysis

CASTILLO, Juan A., Dr.
Department of Analytical Chemistry
University of Zaragoza
Plaza S. Francisco
50009 Zaragoza
Phone: +34 976 761289
Fax: +34 976 761292
E-Mail: jcastilo@porta.unizar.es

Private address:
Francisco Vitoria, 20-22
50008 Zaragoza

Domains of application:
Analytical spectroscopy; optochemical sensors; electrochemical sensors; analytical instrumentation

Applied methods:
Speciation; solid sampling; biochemical analysis

Membership:
ACS; RSC; Spanish Analytical Chemistry Society

CERDÁ, Víctor, Full Prof.
Universitat de les Illes Balears
Ctra. De Valldemosa, s/n
07071 Palma de Mallorca
Phone: +34 971 173261
Fax: +34 971 173426
E-Mail: vcerda@p01.uib.es

Private address:
Capitán Salom, 30
08004 Palma de Mallorca

Domains of application:
Laboratory automation; environmental analysis

Applied methods:
Sequential injection analysis; flow injection analysis; atomic and molecular spectroscopy; chromatographic techniques; electroanalytical techniques

Membership:
Association of Environmental Sciences and Techniques; SEQA; International Association of Environmental Analytical Chemistry; Associación Nacional de Químicos de España

---

CERVERA, M. Luisa, Dr.
Department of Analytical Chemistry
University of Valencia
Dr. Moliner, 50
46100 Burjasot
Phone: +34 96 3983158
Fax: +34 96 3864436
E-Mail: m.luisa.cervera@uv.es

Domains of application:
Atomic absorption spectroscopy; hydride generation; atomic emission; molecular spectroscopy; FIA; microwave digestion and distillation; metal determination and speciation; clean analytical methods; pharmaceuticals determination

Applied methods:
AAS; AE; metal speciation in waters, foods; biological and environmental samples; different digestion procedures of samples; pharmaceutical analysis; FTIR; UV-vis; solid phase extraction and preconcentration of metals

---

CHICHARRO, Manuel, Dr.
Departamento de Química Analítica
Universidad Autónoma de Madrid
Facultad de Ciencias
28049 Madrid
Phone: +34 91 3974041
Fax: +34 91 3974931
E-Mail: manuel.chicharro@uam.es

Domains of application:
Analytical chemistry; electroanalytical chemistry; capillary electrophoresis and HPLC

Applied methods:
Methods of determination in antidoping control

CID, Alfonso, Dr.
Departamento de Química Analítica
Facultad de Farmacia

Marqués de Urquijo, s/n
01080 Vitoria
Phone: +34 945 183057

Private address:
PO Box 804 AP
01080 Vitoria

Domains of application:
Analytical chemistry; pharmaceutical analysis; environmental analysis; electroanalysis; speciation

Applied methods:
Electrochemical methods; spectrophotometric methods; electrophoresis

Membership:
Society of Environmental Toxicology and Chemistry

Awards:
Master on Environmental Engineering and Sciences

COSTA GARCÍA, Agustín, Dr.
Departamento de Química Física y Analítica
Universidad de Oviedo
Julián Clavería, 8
33006 Oviedo
Phone: +34 985 103488
Fax: +34 985 103125

Private address:
Meres, 18
33199 Siero

Domains of application:
Trace analysis; biosensors
Applied methods:
Electroanalytical methods: FIA and HPLC

CRUCES BLANCO, Carmen, Dr.
Department of Analytical Chemistry
Faculty of Sciences
Fuentenueva, s/n
18071 Granada
Phone: +34 958 243297
Fax: +34 958 243328
E-Mail: mcruces@goliat.ugr.es

Private address:
La Cortijuela, 12
18008 Granada

Domains of application:
Spectrophotometry; spectrofluorimetry; phosphorimetry; analytical detection techniques; gas chromatography; mass spectrometry; HPLC; absorption detection

Applied methods:
Analytical methods proposed with the techniques indicated above are applied to enviromental (water, pesticides) food (additives, foodstuffs) and pharmaceutical analysis

Awards:
Extraordinary Licentiation Award of te University of Granada, 1985; San Alberto Magno Award to the Doctor Thesis from the National Association of Spanish Chemists, 1987; First Ronals Belcher Memorial Award of the Royal Society of Chemistry, 1986

CUADROS RODRÍGUEZ, Luis, Dr.
Department of Analytical Chemistry
University of Granada
Fuentenueva, s/n
18071 Granada
Phone: +34 958 243297
Fax: +34 958 243328
E-Mail: lcuadros@goliat.ugr.es

Domains of application:
Quality assurance; chemometrics; toxicological food analysis; environmental analysis; pharmaceutical analysis

Applied methods:
Spectrophotometry and luminescence; chromatographic techniques

---

DE LA GUARDIA, Miguel, Dr.
Central Service of Support of Experimental Research
University of Valencia
Dr. Moliner, 50
46100 Valencia
Phone: +34 96 3864838
Fax: +34 96 3864838

Private address:
Lladró y Malli, 25-20
46007 Valencia

Domains of application:
Atomic spectroscopy; analytical molecular spectroscopy; separation sciences; involving in all these fields basic and applied research in sample preparation and sample introduction; automatization; data treatment; on-line detoxification of waste

Applied methods:
Environmental; food and clinical determinations; industrial applications in fuels; ceramics; pharmaceuticals; paints

Membership:
SEQA, Grupo Espectroquímico

DEL CASTILLO, Benito, Dr.
Laboratory of Instrumental Techniques
Complutense University of Madrid
Plaza Ramón y Cajal, s/n
28040 Madrid
Phone: +34 91 3941754
Fax: +34 91 3941705

Domains of application:
Study of new fluorogenic reagents and their application in pharmaceutical analysis; spectrofluorimetric study of the interaction among DNA and drugs; spectrofluorimetric study of the inclusion complexes among; cyclodextrins and sustances with pharmaceutica

Applied methods:
Molecular spectrofluorimetry; HPLC; atomic absorption spectrophotometry; IR

Membership:
ACS; American Society for Photobiology; SEQA; RSC

Awards:
In the Royal Academy of Pharmacy, Institute of Spain

DÍAZ CRUZ, José Manuel, Dr.
Departamento de Química Analítica
Universidad de Barcelona
Avda. Diagonal, 647
08028 Barcelona
Phone: +34 93 4029083
Fax: +34 93 4021233
E-Mail: jmdiaz@zeus.qui.ub.es

Domains of application:
Environmental chemistry; control analysis; fundamentals of analytical chemistry; chemometrics

Applied methods:
Modern polarographic techniques; stripping analysis; application of chemometrics to electroanalysis

---

DÍAZ GARCÍA, Marta Elena, Prof.
Department of Physical and Analytical Chemistry
University of Oviedo
Av. Julián Clavería, 8
33006 Oviedo
Phone: +34 985 103471
Fax: +34 985 103125
E-Mail:

medg@dwarf1.quimica.uniovi.es

Domains of application:
Chromatographic separations (GC, HPLC, CE); molecular spectroscopic analysis (UV-vis, fluor., phosphor.); electroanalytical chemistry (ASV, electrochem-sensors); atomic spectroscopic methods (AA, ICP, MIP, GD)

Applied methods:
Spectrophotometric methods (UV-vis, fluorescence, room temperature phosphorescence); development of optical (bio)chemical sensors; flow injection immunoassays; food analysis; determination of pollutants in aquatic environments

Membership:
RSC; SEQA

Awards:
Premio Jóvenes Investigadores (Le Recherche, Mundo Científico); Premio San Alberto Magno del Colegio de Químicos; Premio Antonio Hidalgo (Perkin Elmer); Premio Mieres del Camino a la Investigación

---

DOMÍNGUEZ, Elena, Dr.
Department of Analytical Chemistry
University of Alcalá
Faculty of Pharmacy
28871 Alcalá de Henares
Phone: +34 91 8854666
Fax: +34 91 8854666
E-Mail:
qadominguez@inqui.alcala.es

Private address:
López de Hoyos, 462
28043 Madrid

Domains of application:
Biosensors: electrocatalysis, bioelectrochemistry, DNA sensors, new electrode materials; flow systems: biodetection, on-line process monitoring

Applied methods:
FIA; FIIA; simultaneous sugar determination (food matrices, on-line, fermentation control); pesticides; phenols (environmental waters); clinical analysis; catecholamines; salicylic acid

Membership:
International Association of Environmental Analytical Chemistry

---

DUÑACH, Jaime, Dr.
Departamento de Química
Universitat Autónoma de Barcelona
08193 Bellaterra

Applied methods:
Laboratory methods of analytical chemistry; spectroscopy UV-vis; fluorescence; chromatography

Membership:
Societat Catalana de Química; SEQA; Spanish Royal Society of Chemistry

---

ESCOBAR, Rosario, Prof.
Department of Analytical Chemistry
University of Seville
Prof. García González, s/n
41012 Sevilla
Phone: +34 95 4557170
Fax: +34 95 4557168

Domains of application:
Atomic and molecular spectroscopy; electroanalytical methods; chromatography (HPLC, GC); elemental analysis

Applied methods:
FI-chemiluminescence analysis; water analysis; food analysis; nucleotides analysis

Membership:
New York Academy of Sciences; American Association for Advancement of Science; Association of Environmental Sciences and Techniques; SEQA

Awards:
Ministerio de Educación y Ciencia (Servicio de Extensión Universitaria e Investigación) for the programme "An Educational Project of Lab Classes for Environmental Study", 1986

ESTEBAN, Miquel, Dr.
Department of Analytical Chemistry
University of Barcelona
Avda. Diagonal, 647
08028 Barcelona
Phone: +34 93 4021277
Fax: +34 93 4021233
E-Mail: miquel@zeus.qui.ub.es

Private address:
Floridablanca, 133
08011 Barcelona

Domains of application:
Environmental analytical chemistry; control analysis; fundamental analytical chemistry: equilibria in solution chemistry; chemometrics: fundamentals and applications

Applied methods:
Modern polarographic techniques applied to metal speciation; stripping analysis applied to metal speciation; application of chemometric techniques to electroanalytical chemistry

Membership:
SEQA; Societat Catalana de Química

ESTEVE ROMERO, Josep, Dr.
Área de Química Analítica
University of Jaume I
Ctra. Borriol, s/n
12080 Castelló
Phone: +34 964 345700
Fax: +34 964 345654
E-Mail: estevej@nuvol.uji.es

Domains of application:
Pharmaceuticals and clinical chemistry; environmental analysis

Applied methods:
Chromatography; capillary electrophoresis; spectrophotometry-fluorescence

---

FERNÁNDEZ GUTIÉRREZ, Alberto, Dr.
Faculty of Sciences
University of Granada
Fuentenueva, s/n
18071 Granada
Phone: +34 958 243297

Domains of application:
Molecular spectroscopy (fluorescence, phosphorescence) and electroanalytical techniques

Applied methods:
Environmental analysis

FERNÁNDEZ RECAMALES, M. Ángeles, Dr.
Departamento de Química y Ciencia de los Materiales
Universidad de Huelva
Escuela Politécnica Superior
21819 Palos de la Frontera
Phone: +34 959 530213
Fax: +34 959 350311
E-Mail: reca@uhu.es

Domains of application:
Speciation of organometallic ompounds in environmental samples (Sn, As, Se, Hg); mobility of heavy metals from sediments (sequential extractions schemes); environmental

reference materials

Applied methods:
Atomic spectrometry (AAS, GFAAS, ICP-MS); chromatography (GC, HPLC); coupled methods (GC-FPD, GC-MS, HPLC-UV-HG-AAS)

FERNÁNDEZ ROMERO, Juan Manuel, Dr.
Department of Analytical Chemistry
University of Córdoba
Faculty of Sciences
14004 Córdoba
Phone: +34 957 218614
Fax: +34 957 218606
E-Mail: qa2feroj@lucano.uco.es

Domains of application:
Automation of analytical methods; flow-through (bio)chemical sensors; immunoassay; liquid chromatography

Applied methods:
Continuous solid sample treatments

FERNÁNDEZ SÁNCHEZ, M. Luisa, Dr.
University of Oviedo
Julián Clavería, 8
33006 Oviedo
Phone: +34 985 103485
Fax: +34 985 103125

Domains of application:
Analytical chemistry and physical chemistry

FLORIDO, Antonio, Dr.
Department of Chemical Engineering
Politechnical University of Catalunya
Avda. Diagonal, 647
08028 Barcelona
Phone: +34 93 4016554
Fax: +34 93 4015814
E-Mail: florido@eq.upl.es

Domains of application:
Industrial waste treatment; chemical analysis and real-time monitoring

Applied methods:
Potentiometric sensors; new ionophores for ion selective electrodes; real time monitoring of pilot plants for the recovery of heavy metals

Membership:
SEQA

FRÍGOLA CÁNDUES, Ana, Dr.
Facultad de Farmacia
Universidad de Valencia
Av. Vicent Andrés Estellés, s/n
46100 Burjasot
Phone: +34 96 3864955
Fax: +34 96 3864954

Domains of application:
Análisis de alimentos: determinación de elementos traza esenciales y tóxicos, determinación de vitaminas

Applied methods:
Espectrofotometría de absorción atómica: llama y cámara; cromatografía líquida; espectrofluorimetría; espectrofotometría UV-visible; polarografía

---

GALCERÁN, M. Teresa, Dr.
Department of Analytical Chemistry
University of Barcelona
Faculty of Chemistry
08028 Barcelona
Phone: +34 93 4021375
Fax: +34 93 4021233
E-Mail:
galceran@zeus.zeus.qui.ub.es

Domains of application:
Analytical chemistry; environmental

analytical chemistry; separation techniques; chromatography; MS

Membership:
IUPAC; International Association of Environmental Analytical Chemistry

Awards:
Lavoasier Medal, Euroanalysis VI, París

GALLARTA, Félix, Dr.
University de La Rioja
Obispo Bustamente, 3
26001 Logroño
Phone: +34 941 244811
Fax: +34 941 259431

Domains of application:
Teaching; investigation

GALLEGO FERNÁNDEZ, Mercedes, Dr.
Department of Analytical Chemistry
University of Córdoba
Edificio C-3 Anexo. Campus de Rabanales
14071 Córdoba
Phone: +34 957 211066
Fax: +34 957 218616
E-Mail: qa1sirom@lucano.uco.es

Domains of application:
Spectrophotometry; AAS;AES; continuous separation techniques; slurry techniques; automation of sample pretreatment; continuous microwave digestion

Applied methods:
Foods; clinical and biomedical analysis; doping control; environmental; enantiomeric discrimination

Awards:
Separation Techniques Acceeist from Hewlett Packard, Madrid, 1995

GALLO HERMOSA, Blanca, Dr.
Departamento de Química Analítica
Facultad de Ciencias
Apartado 644
48080 Bilbao
Phone: +34 944 648800
Fax: +34 944 648500
E-Mail: qapgaheb@lg.ehu.es

Domains of application:
Biopharmaceutical analysis; food analysis; environmental analysis

Applied methods:
Chromatography; spectroscopy; electrochemistry

GARCÍA ÀLVARES-COQUE, María Celia, Dr.
Department of Analytical Chemistry
University of Valencia
Dr. Moliner, 50
46100 Burjasot
Phone: +34 96 3983003
Fax: +34 96 3864436
E-Mail: celia.garcia@uv.es

Domains of application:
Spectroscopy; chromatography-HPLC; chemometrics; capillary electrophoresis

Membership:
Sociedad Española de Química Analítica

Awards:
National Chemistry Licenciature Award, 1976; Chemistry Licenciature Award Univ. Valencia, 1978; Chemistry PhD Award Univ. Valencia, 1980; Cañada Blanch-Phd Award, 1983

GARCÍA ASUERO, Aguntín, Dr.
Department of Analytical Chemistry
University of Seville
Prof. García González, s/n

Spain

41012 Sevilla
Phone: +34 95 4556746
Fax: +34 95 4556749
E-Mail: asuero@fafar.us.es

Private address:
Sta. M. del Rocío, 4
41008 Seville

Domains of application:
Simultaneous equilibris (multi-step); non aqueous medium; chemometrics: parameter estimation

Applied methods:
Potentiometric method; spectrophotometric methods; HPLC methods; voltammetric methods

Membership:
RSC

Awards:
Academician of Latin-American Academy of Pharmacy

GARCÍA CAMPAÑA, A.M., Dr.
Department of Analytical Chemistry
University of Granada
Av. Fuentenueva, s/n
18071 Granada
Phone: +34 958 243297
Fax: +34 958 243328
E-Mail: amgarcia@goliat.ugr.es

Domains of application:
Development of spectrophotometric and fluorimetric methods; applications of foods and toxicological problems; applications of chemometrics tools and development of models; environmental analysis; pharmaceutical analysis; quality assurance

Applied methods:
Spectrophotometry; luminescence; chromatographic techniques

GARCÍA DE TORRES, Amparo, Dr.
Department of Analytical Chemistry
Faculty of Sciences
Campus de Teatinos
29071 Málaga
Phone: +34 952 131885
Fax: +34 952 132000
E-Mail: agarciad@uma.es

Domains of application:
ICP emission spectrometry; atomic absorption spectrometry; UV-visible spectrophotometry; flow injection analysis; liquid chromatography; chemometrics; molecular fluorescence spectrometry; environmental analysis; history of the sciences

Applied methods:
Analysis and speciation of metal ions; analysis of pesticides, water, environmental and biological samples

---

GARCÍA GARCÍA, M. Soledad, Dr.
Department of Analytical Chemistry
University of Murcia
Campus de Espinardo
30071 Murcia
Phone: +34 968 307100
Fax: +34 968 364148
E-Mail: msgarcia@fcu.um.es

Private address:
Olof Palme, 3
30009 Murcia

Domains of application:
Ion selective electrodes; optical sensors; piezoelectric sensors; flow injection analysis; ion transfer across the interface between two immiscible electrolyte solutions by cyclic voltammetry; spectrophotometric and spectrofluorimetrics methods

---

GARCÍA MONTELONGO, Francisco J., Dr.
Department of Analytical Chemistry, Nutrition and Food Science
University of La Laguna
38071 La Laguna
Phone: +34 922 635679
Fax: +34 922 635602
E-Mail: fjgarcia@ull.es

Domains of application:
Teaching analytical chemistry; research in environmental pollution

Applied methods:
Gas chromatography (FID, ECD, NPD,mass); HPLC (UV-VIS, fluorimetric); capillary electrophoresis; AAS-flame, GF

Membership:
RSC; Spanish Society for Analytical Chemistry

GARRIGUES, Salvador, Dr.
Department of Analytical Chemistry
University of Valencia
Dr. Moliner, 50
46100 Burjasot
Phone: +34 96 3983158
Fax: +34 96 3864838
E-Mail: salvador.garrigues@uv.es

Private address:
Ramón y Cajal, 29
46100 Meliana

Domains of application:
Analytical molecular spectroscopy; analytical atomic spectrometry; separation sciences; involving in all these fields basic and applied research in sample preparation and sample introduction; automatization; data treatment; on-line detoxification of waste

Applied methods:
Environmental, food and clinical determinations; industrial applications in fuels composition, ceramics, pharmaceutical paints

Membership:
SEQA; Grupo Electroquímico

Awards:
The SEQA Award to Beginner Researchers

GÓMEZ ARIZA, José Luis, Prof.
Departamento de Química y Ciencias de los Materiales
Universidad de Murcia
Escuela Politécnica Superior
21819 Palos de la Frontera
Phone: +34 959 530213
Fax: +34 959 350311
E-Mail: ariza@uhu.es

Domains of application:
Speciation of organometallic compounds, in environmental samples (Sn, As, Se, Hg); mobility of heavy metals from sediments (sequential-extraction schemes); environmental reference materials

Applied methods:
Atomic spectrometry (AAS, GFAAS, ICP-MS); chromatography (GC, HPLC); coupled methods (GC-FPD, GC-MS; HPLC-UV-HG-AAS)

GÓMEZ HENS, Agustina, Dr.
Department of Analytical Chemistry
University of Córdoba
Edificio C-3 Anexo. Campus de Rabanales
14071 Córdoba
Phone: +34 957 218645
Fax: +34 957 218644
E-Mail: qa1gohea@lucano.uco.es

Domains of application:
Molecular and atomic spectroscopy; electroanalysis; chromatography; automatic methods; kinetic methods

Applied methods:
Clinical, environmental, pharmaceutical and food analysis

GONZÁLEZ, Elisa, Dr.
Universidad de Vigo
Lagoas-Marcosende, s/n
36200 Vigo
Phone: +34 986 812300
Fax: +34 986 812382
E-Mail: eromero@uvigo.es

Domains of application:
Electroanalytical techniques for the study and determination of psychotropic drugs in biological fluids; development of new biosensor and application in environmental samples, in food and beverages samples and biological fluids

Applied methods:
Tifluadom, Mataclzepam and Clozapine in urine and serum; phenolic compounds in fruits; organic peroxides in environmental samples, using HRP and ARP amperometric sensors; alcohols in beverages as beer, liquor, wine, using ADH and AOX as enzymes

Membership:
ACS

GONZÁLEZ-CORTÉS, Araceli, Dr.
Department of Analytical Chemistry
Complutense University of Madrid
Faculty of Chemistry
28040 Madrid
Phone: +34 91 3944315
Fax: +34 91 3944329

Domains of application:
Analytical chemistry

Applied methods:
Electroanalytical chemistry; electrothermal sensor; electrochemical detection

---

GUILLÉN, D.A., Prof., Dr.
Analytical Chemistry Department
University of Cádiz
Campus Universitario Río San Pedro
11510 Puerto Real
Phone: +34 956 830251
Fax: +34 956 834924
E-Mail: dominico.guillen@uca.es

Domains of application:
Wine analysis; vinegard analysis; brandy analysis; phenolic compounds; organic acids; aroma; browning; chemometric; automation

Applied methods:
HPLC; MS; GC; HPLC-MS; EC; GC-MS; GC-LV; sample preparation

Membership:
Group of Polyphenols

HARDISSON, Arturo, Dr.
Department of Pediatics, Gynecology, Public Health and Toxicology
University of La Laguna
Faculty of Medicine
La Laguna
Phone: +34 922 603503
Fax: +34 922 603407
Private address:
Avda. Lucas Vega, 22
La Laguna

Domains of application:
AAS metals in toxicology; HPLC food additives; ion selective electrodes, fluoride, nitrates and nitrites in beverages and foods; food chemistry; food toxicology

Applied methods:
Chromatographic methods; atomic absorption spectroscopy; inductively coupled plasma

Membership:
The Institute of Food Technologists

HERNÁNCEZ-ARTIGA, María Purificación, Dr.
Department of Analytical Chemistry
University of Cádiz
Apartado 40
11510 Puerto Real
Phone: +34 956 830251
Fax: +34 956 834924

Domains of application:
Analytical chemistry

HERNÁNDEZ, Félix, Dr.
Department of Experimental Sciences
University Jaime I
Campus de Borriol
12080 Castellón
Phone: +34 964 345724
Fax: +34 964 345654
E-Mail: hernandf@exp.uji.es

Private address:
Tarragona, 10
12003 Castellón

Domains of application:
Pesticide residue analysis; coupled chromatographic techniques (LC-LC, LC-GC); trace metal analysis

Applied methods:
HPLC; GC: LC-LC; LC-GC; AAS

Awards:
Premio "Técnicas Analíticas de Separación", Hewlett-Packard, 8as JAI, Barcelona

---

HERNÁNDEZ-CÓRDOBA, Mnauel, Dr.
Department of Analytical Chemistry
University of Murcia
Faculty of Chemistry
30071 Murcia
Phone: +34 968 307100
Fax: +34 968 364148
E-Mail: hcordoba@fcu.um.es

Domains of application:
Atomic absorption spectrometry; ETAAS; FAAS; automation; continuous flow analysis; expert systems

---

HIDALGO-HIDALGO DE CISNEROS, José L., Dr.
Department of Analytical Chemistry
University of Cádiz
Apartado 40
11510 Puerto Real
Phone: +34 956 830251
Fax: +34 956 834924
E-Mail: jluis.hidalgo@uca.es

Domains of application:
Electroanalytical techniques; environmental analysis; chemical sensors

Applied methods:
Water analysis; pollutant analysis; drug analysis

JIMÉNEZ-LINARES, c., Dr.
Department of Analytical Chemistry
University of Granada
Avda. Fuentenueva, s/n
18071 Granada
Phone: +34 958 243327
Fax: +34 958 243328
E-Mail: clinares@golist.ugr.es
Domains of application:
Developments of spectrophotometric and fluorimetric methods; applications on foods and toxicological problems; applications of chemometrics foods and development of model

---

LASERNA, J. Javier, Dr.
Department of Analytical Chemistry
University of Málaga

29071 Málaga
Phone: +34 952 131881
Fax: +34 952 132000
E-Mail: laserna@uma.es

Domains of application:
Analytical chemistry

Applied methods:
Lasers in chemical analysis; laser processing and analysis of materials

Membership:
IUPAC

LEMUS GALLEGO, José María, Dr.
Department of Analytical Chemistry
Faculty of Sciences
Campus Universitario, s/n
13003 Ciudad Real
Phone: +34 926 295300
Fax: +34 926 295318
E-Mail: jmlemus@qata-cr.uclm.es

Domains of application:
Drugs analysis; environmental analysis

LEÓN-GONZÁLEZ, M.E., Dr.
Departamento de Química Analítica
Universidad Complutense de Madrid
28040 Madrid
Phone: +34 91 3944319
Fax: +34 91 3944329
E-Mail:
leongon@eucmax.sim.ucm.es
Domains of application:
Chromatography; mass spectrometry; electrochemistry; sensors; ICP; AA; environmental; drugs; food additives

---

LLAURADÓ, Montserrat, Dr.
Department of Analytical Chemistry
University of Barcelona
Diagonal, 647
08028 Barcelona
Phone: +34 93 4029083
Fax: +34 93 4021233
E-Mail: montse@zeus.qui.ub.es

Domains of application:
Environmental radioactivity

Applied methods:
High resolution gamma spectrometry; liquid scintillation detection

Membership:
International Union of Radioecology

LÓPEZ MOLINERO, Ángel, Dr.
Department of Analytical Chemistry
University of Zaragoza
Pedro Cerbuna, 13
50009 Zaragoza
Phone: +34 976 761174
Fax: +34 976 761292
E-Mail: anlopez@posta.unizar.es

Private address:
Violante de Hungría, 7
50009 Zaragoza

Domains of application:
Teaching analytical chemistry: university level; investigation on analytical chemistry: spectrometry, sensors, chromatography, chemometrics

Applied methods:
Atomic emission spectroscopy in plasmas-ICP, MIP; pattern recognition: principal components analysis, artificial neuronal networks

---

LÓPEZ-CUETO, Guillermo, Dr.
Department of Analytical Chemistry
University of Alicante
Ctra. De San Vicente
03080 Alicante
Phone: +34 965 903527
Fax: +34 965 903464

Domains of application:
Education and research in analytical chemistry

Applied methods:
Atomic spectroscopy; electroanalysis; kinetic analysis; food science

---

LÓPEZ-GARCÍA, Ignacio, Dr.
Department of Analytical Chemistry
University of Murcia
Faculty of Chemistry
30071 Murcia
Phone: +34 968 307100
Fax: +34 968 364148
E-Mail: ilgarcia@fm.um.es

Domains of application:
Atomic absorption spectrometry; ETAAS; FAAS; continuous flow analysis

---

LÓPEZ-SÁNCHEZ, José Fermín, Dr.
Department Química Analítica
Universitat de Barcelona
Avda. Diagonal, 647
08028 Barcelona
Phone: +34 93 4029083
Fax: +34 93 4021233
E-Mail: fermin@zeus.qui.ub.es

Domains of application:
Analytical chemistry education; working group in trace metal speciation in environmental matrices
Applied methods:
Separation techniques; GC; HPLC; CZE; atomic absorption and emission techniques; FAAS; GFAAS; ICP-OES; ICP-MS

LORENZO-ABAD, Encarnación, Dr.
Universidad Autónoma de Madrid
Ctra. Colmenar, Km. 15
28049 Madrid
Phone: +34 91 3974488
Fax: +34 91 3974931
E-Mail: encarna.lorenzo@uam.es

Domains of application:
Design and development of amperometric and microgravimetric biosensors

Applied methods:
Electrochemical and quartz crystal microbalance

Awards:
Comunidad Autónoma de Madrid, Senior Fellow, 1990; Ministerio de Educación y Ciencia, Senior Fellow, 1991; Nato Scientific Program, Senior Fellow, 1993

LORES, Marta, Dr.
Department of Analytical Chemistry
University of Santiago de Compostela
Avda. de las Ciencias, s/n
15706 Santiago de Compostela
Phone: +34 981 547141
Fax: +34 981 547141
E-Mail: qnmlores@usc.es

Domains of application:
Environmental analytical chemistry; food and beverages analytical chemistry; instrument development; chemometrics

Applied methods:
LC (DAD-FL-EQ); micro-LC; GC-MS; GC-AED; GC-FTIR; CZE; SFE; microwave extraction

---

LUQUE DE CASTRO, María Dolores, Dr.
Department of Analytical Chemistry
University of Córdoba
Edificio C-3 Anexo. Campus de Rabanales
14071 Córdoba
Phone: +34 957 218615

Fax: +34 957 218615
E-Mail: qa1lucam@uco.es

Private address:
Avda. Conde Vallellano, 1
14004 Córdoba

Domains of application:
Automation of analytical methods; kinetic methods; validation; quality control; low-through (bio)chemical sensors; immunoassay

Applied methods:
Robotic; continuous; solid sample treatments

MAÑES-VINUESA, Jordi, Dr.
Department of Medicine Preventive
Laboratory of Trace Chemistry and Toxicology
Av. Vicent Andrés Estellés, s/n
46100 Burjasot
Phone: +34 96 3864288
Fax: +34 96 3864954
E-Mail: jordi.manes@uv.es

Domains of application:
Analysis of pesticide and related compounds in food and waters

Applied methods:
GC, LS, SPE

MARINA-ALEGRE, María Luisa, Dr.
Departamento de Química Analítica
Universidad de Alcalá de Henares
Ctra. Madrid-Barcelona, Km. 33,600
28871 Alcalá de Henares
Phone: +34 91 8854935
Fax: +34 91 8854971
Domains of application:
Food, drugs, waters (natural waters and waste waters); soils

Applied methods:
HPLC; CE; UV-VIS spectrophotometry; atomic absorption spectroscopy; GC

MARINÉ-FONT, Abel, Dr.
Department of Human Physiological Science and Nutrition
University of Barcelona
Av. Juan XXIII, s/n
08028 Barcelona
Phone: +34 93 4024509
Fax: +34 93 4021896

Private address:
Providencia, 129
08912 Badalona

Domains of application:
Food chemistry; food science; food analysis; nutrition

Applied methods:
HPLC; GC; other general and instrumental methods

Membership:
International Life Sciences Institute

MARTÍN, Antonio, Dr.
Departamento de Química y Edafología
Universidad de Navarra
Irunlarrea, s/n
31080 Pamplona
Phone: +34 948 425600
Fax: +34 948 425649

Domains of application:
Trace element analytical chemistry; environmental chemistry; deteriorization and conservation of building materials in monuments

Applied methods:
Chemical methods of analysis; instrumental method of chemical analysis

---

MARTÍNEZ, M. del Pilar, Dr.
Faculty of Sciences

University of Cádiz
Polígono Río San Pedro
11510 Puerto Real
Phone: +34 956 830251
Fax: +34 956 834924

Domains of application:
Determination of heavy metals in sediments and waters

Applied methods:
Absorption atomic, spectrophotometry UV, ICP

MARTÍNEZ-VIDAL, José Luis, Dr.
University of Almería
Ctra. Sacramento, s/n
04120 La Coruña
Phone: +34 950 215037
Fax: +34 950 215073
E-Mail: jlmartin@ualm.es

Domains of application:
Analytical chemistry

Applied methods:
Determination of pesticides in air, water, vegetables and fruits

---

MASPOCH, Santiago, Dr.
Departamento de Química
Universidad Autónoma de Barcelona
Facultad de Ciencias
08193 Bellaterra
Phone: +34 93 5811011
Fax: +34 93 5812477
E-Mail: iqan7@blues.uab.es

Domains of application:
Analytical, inorganic, organic and physical chemistry; molecular spectroscopy: NIR, UV-vis, IR-FT, fluorescence, applied chemometrics

Applied methods:
Pharmaceutical analysis; petrochemical analysis; textiles

Membership:
Society for Applied Spectroscopy

MAURI AUCEJO, Adela R., Dr.
Universidad Valencia

MAYNAR-MARIÑO, Juan I., Dr.
Departamento de Química
Analítica y Electroquímica
Universidad de Extremadura
Av. Elvás, s/n
06071 Badajoz
Phone: +34 924 289392
Fax: +34 924 274244
E-Mail: jimaynar@ba.unex.es

Domains of application:
Trace organic in environmental and biological samples; doping analysis; HPLC; AAS; GC-MS

Applied methods:
Doping analysis

MÉNDEZ-GONZÁLEZ, Javier, Dr.
Instituto de Salud Carlos III
Centro Nacional de Sanidad Ambiental
Ctra. Majadahonda-Pozuelo, Km. 2
28220 Majadahonda
Phone: +34 91 5097987
Fax: +34 91 5097927
E-Mail: j.mendez@isiii.es

Domains of application:
Physical and chemical behaviour of atmospheric contaminants; inorganic and organic contaminants in water, wastewaters and sewage; environmental toxicity evaluation on some biological indicators perteining to ecosystems aquatic and earthely

Applied methods:
GEC (VE); EPA; OCDE; UNE; standard methods (APHA; WPLF); ISO

MENÉNDEZ, Alberto, Prof.
Departamento de Química Física y Analítica
Universidad de Oviedo
Julián Clavería, 8
33006 Oviedo
Phone: +34 985 103481
Fax: +34 985 103125
Domains of application:
Química Física y Analítica

Awards:
San Alberto Magno (del Colegio de Químicos de Asturias)

MILLÁN, Esmeralda, Dr.
University of the Basque Country
Manuel de Lardizábal, 3
20080 San Sebastián
Phone: +34 943 216600
Fax: +34 943 212236
E-Mail: qaqmimae@sq.ehu.es

Domains of application:
Heavy metals in environmental samples

Applied methods:
FAAS; EAAS

MIR, José M., Dr.
Department of Analytical Chemistry
Faculty of Sciences
Campus Plaza San Francisco
50009 Zaragoza
Phone: +34 976 761176
Fax: +34 976 761292
E-Mail: jmmir@posta.unizar.es

Domains of application:
Analytical spectroscopy; speciation; hyphenated techniques; GFAAS
Applied methods:
Environmental fields; biotechnology

Membership:
SEQA

MIRANDA-ORDIERES, Arturo, Dr.
Departamento de Química Física y Analítica
Universidad de Oviedo
Julián Clavería, 8
33006 Oviedo
Phone: +34 985 103513
Fax: +34 985 103125
E-Mail: jmiranda@jet.es

Domains of application:
Physical and analytical chemistry, including quantum chemistry, kinetics, spectroscopy, electrochemistry, chromatography and analytical applications to environmental, biomedical and food sciences

Applied methods:
Liquid chromatography; electrochemistry; biosensors

MORENO, Bernardo, Dr.
Departamento de Química Analítica
Universidad de Salamanca
Plaza de la Merced, s/n
37008 Salamanca
Phone: +34 923 294483
Fax: +34 923 294574
E-Mail: bmc@gugu.usal.es

Domains of application:
Analytical chemistry

Applied methods:
Separations (chromatographic, electrophoresis); FI, electronic nose (neural computer sciences); electroanalytical methods; optics methods

---

MORENO-BONDI, María Cruz, Dr.
Department of Analytical Chemistry
Complutense University
Avda. Complutense
28040 Madrid
Phone: +34 91 3944368

Fax: +34 91 3944329
E-Mail:
mcmbondi@eucmax.sim.ucm.es

Domains of application:
Research and teaching in analytical chemistry

Applied methods:
optical methods of analysis; research on optical fiber sensors

Membership:
Spanish Society of Analytical Chemistry; Spectrochemical Group of the Real Sociedad Española de Química

Awards:
Young Researchers Award of the Spanish Society of Analytical Chemistry

---

MURILLO PULGARÍN, José Antonio, Dr.
Department of Analytical Chemistry and Food Technology
Faculty of Sciences
Av. Camilo José Cela, 10
13071 Ciudad Real
Phone: +34 926 295325
Fax: +34 926 295318
E-Mail: jamurill@qata-cr.uclm.es

Domains of application:
Drugs analysis; environmental analysis

Applied methods:
Drugs analysis; environmental analysis

MURILLO, José A., Dr.
Departamento de Química Analítica y Tecnología de los Alimentos
Universidad de Castilla-La Mancha
Av. Camilo José Cela, 10
13071 Ciudad Real
Phone: +34 926 295325
Fax: +34 926 295318
E-Mail: jamurillo@qata-cr.uclm.es

Domains of application:
Educational; research

Applied methods:
Drugs analysis; water analysis

NAVALÓN, Alberto, Dr.
Department of Analytical Chemistry
University of Granada
Av. Fuentenueva, s/n
18071 Granada
Phone: +34 958 243399
Fax: +34 958 243328
E-Mail: anavalon@goliat.ugr.es

Domains of application:
Solid phase spectrometry; spectrofluorimetry; phosphorimetry; gas chromatography; mass spectrometry; atomic absorption spectroscopy; voltammetry

Applied methods:
Environmental analysis; pharmaceutical analysis; food and cosmetic analysis

---

NERIN, Cristina, Dr.
Centro Politécnico Superior
Universidad de Zaragoza
María de Luna, 3
50015 Zaragoza
Phone: +34 976 761873
Fax: +34 976 761861
E-Mail: cnerin@posta.unizar.es

Private address:
Paseo de las Damas, 5
50008 Zaragoza

Domains of application:
Teaching analytical chemistry; research and development in the field of environmental analytical chemistry as well as packaging

Spain

materials in contact with food (migration aspects, virgin polymers and recycled plastics: analytical problems)

Applied methods:
GC-FID; GC-ECD; GC-MS; HPLC-UV; HPLC-fluorescence; AAS; GFAAS; HGAAS; purge and trap method coupled to GC-MS; SFE (CO2) + GC-MS/GC-ECD; SPME + GC-MS/HPLC-UV

Membership:
SEQA; International Association of Environmental and Analytical Chemistry; Association of Environmental Science and Techniques; Red Iberoamericana de Envase y Embalaje

Awards:
1st Price to the Best Research Work in the International Symposium on Analytical Methodology in the Environmental Field, Bilbao, 1996; 1st Price of Environment, Dip. Gral. Aragón, 1990

ORBE-PAYÁ, Ignacio, Dr.
Department of Analytical Chemistry
University of Granada
Av. Fuentenueva, s/n
18071 Granada
Phone: +34 958 243388
Fax: +34 958 243328
E-Mail: idorbe@goliat.ugr.es

Domains of application:
Solid phase spectrometry; spectrofluorimetry

Applied methods:
Environmental analysis; pharmaceutical analysis; food and cosmetic analysis

---

ORTUÑO-SÁNCHEZ-PEDREÑO, Joaquín Ángel, Dr.
Department of Analytical Chemistry
University of Murcia
Campus de Espinardo
30071 Murcia
Phone: +34 968 307100
Fax: +34 968 364148
E-Mail: jortuno@fcu.um.es

Domains of application:
Ion selective electrodes; optical sensors; piezoelectric sensors; flow injection analysis; ion transfer across the interface between two immiscible electrolyte solutions by cyclic voltammetry; spectrophotometric and spectrofluorimetric methods

---

PABLOS, Fernando, Dr.
Department of Analytical Chemistry
University of Seville
Prof. García González, s/n
41012 Sevilla
Phone: +34 95 4557173
Fax: +34 95 4557168
E-Mail: ipablos@cica.es
Domains of application:
Environmental analysis; food analysis; patterns recognition

Applied methods:
Gas chromatography; HPLC; atomic spectroscopy

PALACIOS-CORVILLO, María A., Dr.
Departamento de Química Analítica
Universidad Complutense de Madrid
Ciudad Universitaria
28040 Madrid
Phone: +34 91 3944369
Fax: +34 91 3944329

Private address:
Buen Suceso, 13
28008 Madrid

Domains of application:
Analytical chemistry

Applied methods:
Environmental chemistry

PARDO, Rafael, Dr.
Departamento de Química Analítica
Universidad de Valladolid
Prado de la Magdalena, s/n
47005 Valladolid
Phone: +34 983 423531
Fax: +34 983 423013
E-Mail: rpardo@wamba.cpd.uva.es

Domains of application:
Heavy metal determination; environmental analysis; food analysis; chemometrics

Applied methods:
Voltammetry; ICP; atomic absorption

PASTOR, Agustín, Dr.
Department of Analytical Chemistry
University of Valencia
Dr. Moliner, 50
46100 Burjasot
Phone: +34 96 3864454
Fax: +34 96 3864436
E-Mail: agustin.pator@uv.es

Domains of application:
Atomic absorption and emission spectrometry; ICP-MS; GC; HPLC; ion chromatography; MS; environmental analytical chemistry; metal speciation; analysis of organic micropollutants

Applied methods:
Environmental, biological and food samples; bioaccumulation of metals; formation of metallothioneins

Membership:
Commission Internationale pour l'Exploration Scientifique de la Mer Mediterranee

---

PEDREÑO-MUÑOZ, María, Dr.
Departamento de Química Analítica
Universidad Complutense de Madrid
Ciudad Universitaria
28040 Madrid
Phone: +34 91 3944368
Fax: +34 91 3944329
E-Mail:
mpedrero@eucmax.sim.ucm.es

Domains of application:
Research and teaching in analytical chemistry

Applied methods:
Electroanalytical; spectroscopical; chromatography

Membership:
ACS; SEQA; RSC

PÉREZ-ARRIBAS, Luis Vicente, Dr.
Departamento de Química Analítica
Universidad Complutense de Madrid
Ciudad Universitaria
28040 Madrid
Phone: +34 91 3944319
Fax: +34 91 3944329

Domains of application:
Environmental analysis; food analysis; drugs

Applied methods:
Chromatography; electroanalysis; mass spectrometry; sensors; ICP; AA

PÉREZ-BENDITO, Dolores, Dr.
Department of Analytical Chemistry
University of Córdoba
Edificio C-3 Anexo. Campus de Rabanales
14071 Córdoba
Phone: +34 957 218644

Fax: +34 957 218644
E-Mail: qa1meobj@uco.es

Private address:
Paseo de la Victoria, 21
14004 Córdoba

Domains of application:
Automation of analytical methods; kinetic methods; trace analysis; FIA; immunoassay; chromatography; spectrometry; luminescence; sensors; quality control; kinetometrics

Applied methods:
Environmental analysis; clinical and pharmacological analysis; food analysis

Membership:
ACS; International Association of Environmental Analytical Chemistry

---

PEREZ-BUSTAMENTE, Juan Antonio, Dr.
Departamento de Química Analítica
Facultad de Ciencias
Apartado 40
11510 Puerto de Santa María
Phone: +34 956 830251
Fax: +34 956 834924

Private address:
Hibiscos, 5
11007 Cádiz

Domains of application:
Chemical sensors; organic reagents; geochemical analysis; electrochemistry; wine analysis; chemometrics; separation techniques

Applied methods:
Mass spectrometry; gas chromatography; HPLC; fluorimetry; VIS/UV spectrometry; electrochemical techniques

Membership:
ACS; RSC; Royal Spanish Chemical Society; Spanish Society of Analytical Chemistry; Spanish Society of History of Science and Technology; Society of History of Alchemy and Chemistry; Working Party History of Chemistry-FECS; Working Party Anal. Chem.-FECS

PINGARRÓN, José M., Dr.
Department of Analytical Chemistry
Complutense University of Madrid
28040 Madrid
Phone: +34 91 3944315
Fax: +34 91 3944329

Domains of application:
Analytical chemistry

Applied methods:
Electroanalytical chemistry; electrochemical (bio)sensors; electrochemical detection

Membership:
ISE; European Society for Electroanalytical Chemistry

PINILLA-GIL, Eduardo, Dr.
Departamento de Química Analítica y Electroquímica
Universidad de Extremadura
Av. Elvás, s/n
06071 Badajoz
Phone: +34 924 289389
Fax: +34 924 274244
E-Mail: epinilla@ba.unex.es

Domains of application:
Trace organics in environmental and biological samples; trace elements in environmental and biological samples; LC-ECD; voltammetry; atomic absorption; gas chromatography with MS detection

Applied methods:
Voltammetric methods for drugs and trace elements in real samples

---

POLO DÍEZ, Luis, Dr.
Universidad Complutense
Facultad de Química
Ciudad Universitaria
28040 Madrid
Phone: +34 91 3944326
Fax: +34 91 3944329

Domains of application:
Analytical chemistry

Applied methods:
Water; food; environmental and clinical analysis by spectroscopy and chromatography

Membership:
Real Sociedad Española de Química

PRADA RODRÍGUEZ, Darío, Dr.
Department of Analytical Chemistry
University of A Coruña
Campus de Zapateira, s/n
15071 A Coruña
Phone: +34 981 167000
Fax: +34 981 167065
E-Mail: dariop@udc.es

Domains of application:
Environmental studies; applied spectroscopy; applied chromatography; applied chemometrics

Applied methods:
UV-VIS; FT-MIR; FAAS; GFAAS; mass chromatography (GC, HPLC)

RAMIS RAMOS, Guillermo, Dr.
University of Valencia
Dr. Moliner, 50
46184 Valencia
Phone: +34 96 3983003
Fax: +34 96 3864436
E-Mail: guillermo.ramis@uv.us

Domains of application:
Analytical spectroscopy; capillary electrophoresis; HPLC-liquid chromatography; chemometrics; automation

Membership:
SEQA

RAMIS RAMOS, Guillermo, Dr.
Department of Analytical Chemistry
University of Valencia
Dr. Moliner, 50
46100 Burjasot
Phone: +34 96 3863003
Fax: +34 96 3864436
E-Mail: guillermo.ramis@uv.es

Domains of application:
Spectroscopy; chromatography-HPLC; capillary electrophoresis; chemometrics

Membership:
SEQA; Association of Environmental Sciences and Techniques; Grupo Electroquímico de la Real Sociedad Española de Química

---

RAURET DALMAU, Gemma, Dr.
Department of Analytical Chemistry
University of Barcelona
Marti i Franqués 1
08028 Barcelona
Phone: +34 93 4021278
Fax: +34 93 4021233
E-Mail: rauret@zeus.qui.ub.es

Domains of application:
Environmental Analytical Chemistry

RÍOS, Ángel, Dr.
Department of Analytical Chemistry
University of Córdoba
Edificio C-3 Anexo. Campus de Rabanales
14071 Córdoba
Phone: +34 957 211066
Fax: +34 957 218616

E-Mail: qa1ricaa@uco.es

Domains of application:
Analytical chemistry; automatic methods of analysis; analytical quality systems; FIA; education in analytical chemistry

Applied methods:
Automation and validation of methods in the fields of environmental and food/agricultural analysis

Membership:
SEQA, SEA

Awards:
Young Researchers from Spanish Analytical Society

RIUS, F. Xavier, Dr.
Universitat Rovira i Virgili
Pl. Imperial Tarraco, 1
43005 Tarragona
Phone: +34 977 559562
Fax: +34 977 539563
E-Mail: rius@quimica.urv.es

Domains of application:
Chemistry (physical, polysters, coordination compunds, organic synthesis, biochemistry); chemometrics

Applied methods:
Separation techniques (HPLC, GC, CE); molecular spectroscopy; flow methods; chemometrics

---

RODRÍGUEZ, Juana, Dr.
Department of Analytical Chemistry
Faculty of Sciences
Campus Universitario, s/n
13003 Ciudad Real
Phone: +34 926 295300
Fax: +34 926 295318
E-Mail: jrflores@qata-cr.uclm.es

Domains of application:
Dyes, hormones, drugs; HPLC; spectrophotometry; electrochemistry; drugs analysis; environmental analysis

Awards:
Novel Awards of Analytical Research (1988) (for young researches)

ROMÁN CEBA, M., Dr.
Department of Analytical Chemistry
University of Granada
Faculty of Sciences
18071 Granada
Phone: +34 958 243297
Fax: +34 958 243328

Domains of application:
Development of spectrophotometric and spectrofluorimetric methods

Applied methods:
Application on foods and toxicological problems; application of chemometric tools and development models

---

ROSES, Martí, Dr.
Department of Analytical Chemistry
University of Barcelona
Diagonal, 647
08028 Barcelona
Phone: +34 93 4021796
Fax: +34 93 4021233
E-Mail: marti@zeus.qui.ub.es

Domains of application:
Analytical chemistry; solution chemistry; environmental chemistry

Applied methods:
Chromatography; spectroscopy; electrochemistry; chemometrics

Membership:
ACS; International Group for Correlation Analysis in Chemistry

---

RUBIO BARROSO, Sara, Dr.
Facultad de Ciencias
Universidad Complutense de Madrid
Departamento de Química Analítica
28040 Madrid
Phone: +34 91 3944217
Fax: +34 91 3944329

Private address:
Alburquerque, 19
28010 Madrid

Domains of application:
Medio ambiente; análisis de alimentos

Applied methods:
Métodos cromatográficos; métodos espectroscópicos

RUBIO BRAVO, Soledad, Dr.
Department of Analytical Chemistry
University of Córdoba
Edificio C-3 Anexo. Campus de Rabanales
14071 Córdoba
Phone: +34 957 218643
Fax: +24 957 218644
E-Mail: qa1rubrs@lucano.uco.es

Private address:
Av. Medina Azahara, 49; 14005 Córdoba

Domains of application:
Automation of analytical methods; kinetic methods; trace analysis; chromatography, spectrometry; luminescence; kinetometrics

Applied methods:
Environmental analysis; clinical and pharmacological analysis

Membership:
International Association of Environmental Analytical Chemistry; International Association on Water Quality

---

RUBIO, Roser, Dr.
Departamento de Química Analítica
University of Barcelona
Diagonal, 647
08028 Barcelona
Phone: +34 93 4021283
Fax: +34 93 4021233
E-Mail: rubio@zeus.qui.ub.es

Domains of application:
Trace metals in environmental and biological materials; development of coupled techniques for trace metal speciation; sequential extraction of trace metals in soils and sediments

Membership:
International Association of Environmental Analytical Chemistry; International Water Supply Association

---

SALVADÓ, Victoria, Dr.
University of Girona
Campus Montilivi
17071 Girona
Phone: +34 972 418288
Fax: +34 972 418150
E-Mail: victoria@fc.udg.es

Domains of application:
Separation techniques: solvent extraction, liquid membranes, ion-exchange; sensors (electrodes and optodes); environmental analysis: organic and inorganic contaminants

Applied methods:
AAS; AAS-GF; GC; HPLC; UV-visible spectroscopy

SALVADOR, María Amparo, Dr.
Department of Analytical Chemistry
Faculty of Chemistry
Dr. Moliner, 50
46100 Burjasot
Phone: +34 96 3983175
Fax: +34 96 3864322
E-Mail: salvador@uv.es

Private address:
Peris y Valero, 189
46005 Valencia

Domains of application:
Supercritical fluid chromatography; supercritical fluid extraction; atomic spectrometric techniques; UV-vis spectrometry

Applied methods:
Pharmaceuticals analysis; environmental analysis; clinical analysis

---

SÁNCHEZ MISIEGO, Antonio, Dr.
Facultad de Ciencias
Universidad de Extremadura
Av. Elvás, s/n
06071 Badajoz
Phone: +34 924 289393
Fax: +34 924 274244
E-Mail: amisiego@unex.es

Domains of application:
New electroanalytical and chromatographic strategies in biological sciences

Applied methods:
To evaluation of oligo-elements (essential and toxic) in sea-water, blood, CR liquid; to evaluation and determination of steroid hormones, sphedrines and corticosteroids to doping control

---

SÁNCHEZ PEDREÑO, Concepción, Dr.
Faculty of Chemistry
University of Murcia
Campus de Espinardo
30071 Murcia
Phone: +34 968 307100
Fax: +34 968 364148
E-Mail: spedreno@fcu.um.es

Private address:
Plaza de la Fuensanta, 1
30005 Murcia

Domains of application:
Ion selective electrodes; optical sensors; piezoelectric sensors; FIA; ion transfer across the interface between two immiscible electrolyte solutions by cyclic voltammetry; spectrophotometric and spectrofluorimetric methods

Awards:
Medal of Chemistry. Royal Spanish Society of Chemistry, 1979; Murcia Laurel of Scientific Research, 1972

SÁNCHEZ RASER, Francisco, Dr.
Estación Experimental del Zaidín (CSIC)
Profesor Albareda, 1
15008 Granada
Phone: +34 958 121011
Fax: +34 958 129600
E-Mail: fsrasero@eez.csic.es

Domains of application:
Pesticides/environment; agricultural and earth sciences

Applied methods:
HPLC/GC/MS
Membership:
Collaborative International Pesticides Analytical Council Ltd.; Group of Experts on Pesticide Specifications of the Foos and Agricultural Organization of the United Nations

Awards:
CSIC Award for CIPAC Chairman

SANTANA RODRÍGUEZ, José Juan, Dr.
Department of Chemistry
University of Las Palmas
35017 Las Palmas
Phone: +34 928 452915
Fax: +34 928 452922
E-Mail:
josejuan.santana@quimica.ulpgc.es

Private address:
Gardenia, 17
35212 Telde

Domains of application:
Analytical chemistry; environmental analysis; organic chemistry; natural products; physical chemistry; electrochemistry

Applied methods:
Spectrometry UV-vis; fluorescence and phosphorescence spectrometry; HPLC; application to the determination of organic compounds of environmental interest: PAHs, PCBs, PCDibenzofurans, PCDioxins; Pigments and compounds of biochemical interest: amino acids

Membership:
International Association of Environmental Analytical Chemistry; New York Academy of Sciences

Awards:
Award of PhD with Special Distinction

SANTOS DELGADO, María Jesús, Dr.
Facultad de Ciencias Químicas
Universidad Complutense de Madrid
Departamento de Química Analítica
28040 Madrid
Phone: +34 91 3944323
Fax: +34 91 3944329

Domains of application:
Análisis medioambiental; análisis de alimentos

Applied methods:
Métodos cromatográficos; métodos espectroscópicos

SANZ-MEDEL, Alfredo, Dr.
Department of Analytical Chemistry
University of Oviedo
Julián Clavería, 8
33006 Oviedo
Phone: +34 985 103474
Fax: +34 985 103125
E-Mail:
asm@dwarf1.quimica.uniovi.es

Domains of application:
Analytical chemistry and physical chemistry

Membership:
IUPAC; RSC; Society for Applied Spectroscopy

SILVA RODRÍGUEZ, Manuel, Dr.
Department of Analytical Chemistry
University of Córdoba
Edificio C-3 Anexo. Campus de Rabanales
14071 Córdoba
Phone: +34 957 212099
Fax: +34 957 218614
E-Mail: qa1sirom@uco.es

Domains of application:
Spectrophotometry; luminescence techniques; atomic techniques; chromatography: gas and HPLC; electroanalytical techniques; automatic methods; kinetic methods; chemometrics

Spain

Applied methods:
Environmental, food, clinical and pharmaceutical analysis

Awards:
Graduate Study Award; Prof. Antonio Hidalgo Award

SIMO ALFONSO, Ernesto, Dr.
Department of Analytical Chemistry
University of Valencia
Dr. Moliner, 50
46100 Burjasot
Phone: +34 96 3983176
Fax: +34 96 3864436

Domains of application:
Spectroscopy; HPLC; CE; chemometrics

Membership:
Association of Enviromental Sciences and Techniques

TAULER, Roma, Dr.
University of Barcelona
Diagonal, 647
08028 Barcelona
Phone: +34 93 4021545
Fax: +34 93 4021233
E-Mail: roma@quimio.qui.ub.es

Domains of application:
Analytical chemistry (fundamentals); environmental chemistry; chemometrics; bioanalytical

Applied methods:
Spectroscopy; chemometrics; electroanalytical

Membership:
International Chemometrics Society

Awards:
Premi marti d'Ardenya, Institut d'Estudis Catalans, 1993

TUÑÓN BLANCO, Paulino, Full Prof.
Department of Physical and Analytical Chemistry
University of Oviedo
Julián Clavería, 8
33006 Oviedo
Phone: +34 985 103487
Fax: +34 985 103521
E-Mail: ptb@fluor.quimica.uniovi.es

Domains of application:
Computational and theoretical chemistry; x-ray crystallography; electrochemistry; environmental, chemical, biochemical and industrial analysis; food analysis; sensors

Applied methods:
Electroanalytical methods; chromatography; electrochemical biosensors

Membership:
European Society for Electroanalytical Chemistry

UBIDE, Carlos, Dr.
Facultad de Química
Universidad del Páis Vasco
Apartado 1072
20080 San Sebastian
Phone: +34 943 216600
Fax: +34 943 212236
E-Mail: qapubsec@sq.ehu.es

Domains of application:
Kinetic methods of analysis; chemical kinetics and reaction mechanisms

---

VALCÁRCEL CASES, Miguel, Dr.
Department of Analytical Chemistry
University of Córdoba
Edificio C-3 Anexo. Campus de Rabanales
14071 Córdoba
Phone: +34 957 218616

Fax: +34 957 218616
E-Mail: qa1meobj@uco.es

Private address:
Avda. de la Victoria, 21
14004 Córdoba

Domains of application:
Automation analytical chemistry; theory analytical chemistry; quality assurance on chemical laboratories; sample screening systems; hybridization of instrumental techniques

Membership:
RSC; ACS; Spanish Society of Analytical Chemistry; DAC-FECS

Awards:
F.E. Plakette Award (Austrian Society of Analytical Chemistry); 2nd Price of International Competition on definition of Analytical Chemistry; Chartered Chemist and Fellow RSC; Spanish Awards: Maimonides Price (Junta de Andalucía), Solvay Prize Chem. Sci.

VALIENTE MALMAGRO, Manuel, Dr.
Department de Química Analítica
Universitat Autónoma de Barcelona
Facultad de Ciencias
08193 Bellaterra
Phone: +34 93 5812903
Fax: +34 93 5811985
E-Mail: iqan3@cc.uab.es

Private address:
Passeig Baixador, 87
08190 Mirasol-San Cugat del Vallés

Domains of application:
General analytical chemistry; separation science processes; chemometry; chemical sensors and biosensors

Applied methods:
Membrane separation; preparation and development of new membranes; liquid membranes; hollow fibres; composite activated membranes; chromatographic materials preparation and characterization; development of parametric pumping processes

Membership:
International Committee for Solvent Extraction; European Society of Membrane Science and Technology; American Association for the Advancement of Science; ACS

Awards:
Group of Quality Research: 1993-1995, 1995-1997; Commission of Research and Development, Generalitat de Catalunya, Spain

---

VICENTE ESTÉVEZ, Francisca, Dr.
Departamento de Química Analítica
Facultad de Ciencias
Apartado 644
48080 Bilbao
Phone: +34 944 647700
Fax: +34 944 648500

Domains of application:
Biopharmaceutical analysis; food analysis; environmental analysis

Applied methods:
Chromatography; spectroscopy; electrochemistry

VILCHEZ QUERO, José Luis, Dr.
Department of Analytical Chemistry
University of Granada
Av. Fuentenueva, s/n
18071 Granada
Phone: +34 958 243398
Fax: +34 958 243328
E-Mail: jvilchez@goliat.ugr.es

Private address:
Av. De la Constitución, 18
18012 Granada

Domains of application:
Solid phase spectrofluorimetry; GC-MS

Applied methods:
Environmental analysis; pharmaceutical analysis; water and soil analysis

---

VIÑAS, Pilar, Dr.
Department of Analytical Chemistry
University of Murcia
30071 Murcia
Phone: +34 968 307100
Fax: +34 968 364148
E-Mail: pilarvi@fcu.um.es

Domains of application:
Atomic absorption spectrometry; ETAAS; FAAS; liquid chromatography; automation; continuous flow analysis; expert systems

---

ZAPARDIEL PALENZUELA, Antonio, Dr.
Departamento de Química Analítica y Análisis Instrumental
Universidad Autónoma de Madrid
28049 Cantoblanco
Phone: +34 91 3974937
Fax: +34 91 3974931
E-Mail: antonio.zapardiel@uam.es

Domains of application:
Electroanalytica, Capillary electrophoresis

Sweden

DANIELSSON, Lars-Goran, Dr.
Department of Analytical Chemistry
Royal Institute of Technology
10044 Stockholm
Phone: +46 8 7908215
Fax: +46 8 108425
E-Mail: lgd@analyt.kth.se

Domains of application:
Process analytical chemistry; membrane interfaces

Applied methods:
Spectroscopy (NIR, IR, UV-VIS)

Membership:
IUPAC

Awards:
TAPPI Best Research paper Award 1997

GUSTAVSSON, Ingemar, Dr.
Swedish Institute for Metals Research
Drottning Kristinas vag 48
11428 Stockholm
Phone: +46 8 4404835
Fax: +46 8 4403535
E-Mail: irgn@simr.se

Domains of application:
Wet chemical analysis of steel and iron; ICP-MS analysis for trace element determination in steel, iron, other metals and metal alloys; production of certified reference materials (steel related)

Applied methods:
Spectrophotometry; gravimetry; titration (visual and potentiometric); combustion with IR-detection (HFIR); AAS; ICP-MS

Membership:
ISO-REMCO (Council Committee on Reference Materials); ICO/TC17/SC1 (Committee for Analytical Chemistry for Steel and Iron

---

INGMAN, Folke, Dr.
Royal Institute of Technology
Vaihallavagen 79
10044 Stockholm
Phone: +46 8 7908183
Fax: +46 8 108425
E-Mail: folke@analyt.kth.se

Private address:
Karl Martins v. 13; 8538 Vaxholm

Domains of application:
Process analytical chemistry; spectrometry; multivariate methods; electroanalytical chemistry

Applied methods:
ICP; inorganic MS

Membership:
DAC-FECS; IUPAC

Awards:
Talanta tonis Gordon Memorial Award 1966

JOSEFSSON, Bjorn, Dr.
Department of Analytical Chemistry
Stockholm University
10691 Stockholm
Phone: +46 8 162438
Fax: +46 8 156391
E-Mail: bjosefsson@anchem.su.se

Domains of application:
Trace analysis of organic compounds; environmental research; process analytical chemistry

Applied methods:
Chromatography and electrophoresis; miniaturized separation systems; laser induced fluorescence detection; mass spectrometry; chemometric

---

KARLBERG, Bo, Dr.
Department of Analytical Chemistry
Stockholm University
10691 Stockholm
Phone: +46 8 162437
Fax: +46 8 156391
E-Mail: bkarlberg@anchem.su.se

Domains of application:
Flow injection analysis; microflow systems; capillary electrophoresis; spectroscopy; chemometrics

---

WAHLUND, Karl-Gustav, Dr.
Technical Analytical Chemistry
Center for Chemistry and Chemical Engineering
Lund University
PO Box 124
22100 Lund
Phone: +46 46 2228316
Fax: +46 46 2224525
E-Mail: karl-gustav.wahlund@teknlk.lth.se

Private address:
Sangarevagen 6
22471 Lund

Domains of application:
Process analytical chemistry; imaging of separation processes by LIF/LCD; analysis of chlorinated fatty acids in fish lipids; polymer analysis; starch analysis; polysaccharide analysis; biochemical analysis; micro-chip-based analysis

Applied methods:
Analytical separations; gas chromatography; liquid chromatography (HPLC); field-flow fractionation (FFF); capillary electrophoresis (CE); chemometrics;

light scattering

Membership:
ACS

Switzerland

CAPEROS, Jose R., Dr.
Laboratoire Cantonal
Jehanne-de Hochberg, 5
2001 Neuchatel
Phone: +41 32 8895814
Fax: +41 32 8896274
E-Mail: jose.caperos@acn.etatne.ch

Private address:
Cote, 4
2000 Neuchatel

Domains of application:
Analytical food; food chemistry

Membership:
Societé Francaise de Chimie

SCHWILKOWSKI, Margit, Dr.
Paul Scherrer Institut
5232 Villigen PSI
Phone: +41 56 3104110
Fax: +41 56 3104435
E-Mail: schwikowski@psi.ch

Domains of application:
Trace analyses of inorganic species in various samples, for example in alloys, ceramics, minerals, corrosion products, catalytic materials and in environmental samples like aerosol, snow and ice samples

Applied methods:
ICP-OES; ICP-MS; IC; CE; AAS; NAA; alpha-, beta-, gamma-spectroscopy

Turkey

AKGAY, Mehmet, Prof., Dr.
Cumhuriyet University
58140 Sivas
Phone: +90 346 2261527

Domains of application:
Education/research

APAK, Mustafa Resat, Dr.
Department of Chemistry
Faculty of Engineering
Avcilar Campus
34850 Istanbul
Phone: +90 212 5911996
Fax: +90 212 5911997

Private address:
Ismailapsa Sok. 55
81020 Kosyyolu-Istanbul

Domains of application:
Development of spectrophotometric methods of analysis for trace elements, esp. for environmentally important elements and compounds; solid waste evaluation, treatment, recycling and waste water treatment by unconventional adsorbents

Applied methods:
Atomic and molecular spectrometry (AAS, UV-VIS spectrophotometry, IR-spectrometry); HPLC methods

Membership:
NATO/CMS; IUPAC, New York Academy of Sciences

Awards:
NATO/CCMS Fellowship Award (1988-96); British Council Fellowship (1981)

ARCA, Meral, Dr.
Department of Chemistry

Hacettepe University
06532 Beytepe-Ankara
Phone: +90 312 2352596
Fax: +90 312 2352596
E-Mail: arca@eti.cc.hun.edu.tr

Private address:
Eston Sitesi 14 Bok No. 57
Ankara

Domains of application:
Electroanalytical chemistry; conducting polymers; modified electrodes

Applied methods:
Voltammetric; coulometric; potentiometric; conductimetric; amperometric methods

Awards:
Islamic Development Bank Postdoctroal Merit Scholarship (Nov. 1992 - May 1994). University of Texas at Austin, USA

BECTAY, Sema, Dr.
Department of Chemistry
Hacettepe University
06532 Beytepe-Ankara
Phone: +90 312 2352596
Fax: +90 312 2352596

Private address:
MESA Korn Sitesi Sardunya D Blok
No 31, Ankara

Domains of application:
Atomic spectroscopy; chromatographic methods of analysis

Applied methods:
AAS; HPLC

CANEL, Esin, Dr.
Department of Chemistry
Ankara University
06100 Ankara
Phone: +90 312 2126720
Fax: +90 312 2232395
E-Mail:
ecanel@science.ankara.edu.tr

Private address:
16 Sokak No. 3/11
06490 Ankara

Domains of application:
Determination of equilibria constants; ion-selective electrodes

Applied methods:
Potentiometry; spectrophotometry

DOGAN, H. Mehmet, Prof., Dr.
Department of Chemistry
Hacettepe University
06532 Beytepe-Ankara
Phone: +90 312 2352596
Fax: +90 312 2352596
E-Mail:
mdogan@osiris.chem.hun.edu.tr

Private address:
Borglar Cord. 195
06059 Ankara

Domains of application:
Education of analytical chemistry; trace analysis; preconcentration of trace metals in various matrices; emission spectroscopy; AAS

Applied methods:
Preconcentration of metal ions in various matrices on amer life XAD-resins or on activated carbon

Membership:
Turkish Chemical Society; GDCh; Turkish Spectroscopy Society; Hayseri Yuksekogretim Vakfi; University-Sanayi-Isbirligi Valefi

---

ELCI, Latif, Prof., Dr.
Erciyes University
Art and Sciences Faculty

Chemistry Department
38039 Kayseri
Phone: +90 352 4374901
Fax: +90 352 4374933
E-Mail: elçi@zirve.erciyes.edu.tr

Applied methods:
The method are based on the combination of atomic spectrometry or spectrophotometry with preconcentration techniques; instruments: flame or graphite furnace AAS, DCP-AES; preconcentrations: sorbent extraction, coprecipitations

Membership:
VI National Congress on Spectroscopy; Organizing Committee for Colloquium Spectroscopicum Internationale (CSI)

ELIK, Adil, Dr.
Cumhuriyet University
58140 Sivas
Phone: +90 346 2261527
Fax: +90 346 2263513

Domains of application:
Education/research

ERBATUR, Oktay, Dr.
Fen Edebiyat Fakultesi
Cukirova University
Kimua Bolumu
01330 Adana
Phone: +90 322 3386081
Fax: +90 322 3386070
E-Mail: erbatur@pamuk.cc.cu.edu.tr

Domains of application:
Research on developing techniques for determining toxic organics in environmental matrices; developing electrochemical methods for destroying toxic organics adsorbed on humic acids; cloys and activated carbons

Applied methods:
All sorts of chromatographic techniques coupled to various detectors and electrochemical equipments; spectrophotometers of various types are being used as find analyses in the above type research projects

Membership:
ACS

ERDOGAN, Hasdemir, Assoc. Prof., Dr.
Department of Chemistry
Gazi University
Kimya Bolumu
06500 Ankara
Phone: +90 312 2122900
Fax: +90 312 2122279

FILIK, Hayati, Dr.
Department of Chemistry
Faculty of Engineering
Avvilar Campus
34850 Avcilar- Istanbul
Phone: +90 212 5911998
Fax: +90 212 5911997

Private address:
Merkez Mah. Belediye Cd. Demii St. 8
Avcilar-Istanbul

Domains of application:
Development of spectrophotometric methods of analysis for trace elements, esp. ror environmentally important elements and reducing agents, esp. for biologically important compounds; heavy metals and organic pollutants removal from aqueous salutions

Applied methods:
Molecular and atomic spectrometry

Turkey

GAGLAR, Perihan, Dr.
Department of Chemistry
Hacettepe University
06532 Beytepe-Ankara
Phone: +90 312 2239609
Fax: +90 312 2150678
E-Mail: percag@eti.cc.hun.edu.tr

Private address:
11 Sokak N0 43/16 Bahcelieuler
06490 Ankara

Domains of application:
Trace elements in biological samples; fiber optic chemical sensors and biosensors; electrochemical sensors based on conductive polymers; fiber optic sensors based on polymer smelling; environmental monitoring of heavy metals

Applied methods:
UV-visible and near IR molecular spectroscopy; atomic spectroscopy; molecular spectroscopy using fiber optics; FTIR spectroscopy

Membership:
ACS; IUPAC

Awards:
British Chemical Fellowship (1985); Fulbright Fellowship (1990); Tubitak Fellowship (1985); British Council Research Grant (1995); UNDP Research Grant (1994)

GOKOGLU, Elmas, Assoc. Prof.
Department of Chemistry
Hacettepe University
06532 Beytepe-Ankara
Phone: +90 312 2352596
Fax: +90 312 2352596
E-Mail: gokoglu@eti.cc.hun.edu.tr

Private address:
Medil Sohali 39
Ankara

Domains of application:
Photochemistry; spectroscopic methods; biochemistry; sensors

Applied methods:
Photochemical analysis; luminescence methods; fluorescence; phosphorescence; chemiluminescence

---

GUCER, Seref, Dr.
Department of Chemistry
Science and Arts Faculty
16059 Bursa
Phone: +90 224 4428200
Fax: +90 224 4428022
E-Mail:
gucer@uu20.bim.uludag.edu.tr

Private address:
Kukurdu Caddesi, Dupe Sikak, 3
16080 Cekirge Bursa

Domains of application:
Micro methods and trace analysis; applications in environmental and biological samples

Applied methods:
Gas chromatography; HPLC; ion chromatography; GC-MS; atomic absorption spectrometry (flame graphite); IR and UV/vis spectrometry; potentiometry (ion selective electrodes)

Membership:
IUPAC; American Association for the Advancement of Science; European Microbeam Analysis Society

Awards:
Deutscher Akademische Austauschdienst, 1972; International Seminar (Karlsruhe), 1989-79; NATO Research Grant, 1980-82, 1989-91; National Science Foundation of Belgium, 1987-88; DAAD Research Grant, 1990, 1996;

Visiting Professor in Antwerp University, 1992

HAMAMCI, Candan, Assoc. Prof., Dr.
Department of Chemistry
University of Dicle
21280 Diyarbakir
Phone: +90 412 2488550
Fax: +90 412 2488039
E-Mail: fen@kim.at.tr

Domains of application:
Hydrometallurgy; atomic spectrometry; fuel; air pollution; environmental science; separation technique; extraction

Applied methods:
ICP-AES; AAS; leaching; ion exchange; chromatography; AAW injection; liquid-liquid extraction; wat chemical desulfurization

---

HASAN, Aydin, Assoc. Prof., Dr.
Fen-Edebiyat Fakultesi
Gazi University
Kimya Bolumu
06500 Ankara
Phone: +90 312 2122900
Fax: +90 312 2122279
E-Mail: kimya@quark.fef.gazi.edu.tr

Private address:
Basin Cadoesi Dizgi Sokak, 13
Ankara

Domains of application:
Development of analytical methods; environmental studies and analysis

Applied methods:
Spectroscopic methods; chromatography (gas chromatography and HPLC) and voltammetric methods (cyclic anodic and cathodic stripping voltammetry)

HENDEN, Emir, Prof., Dr.
EGE University
Science Faculty
Bornova
35100 Izmir
Phone: +90 232 3881092
Fax: +90 232 3881036
E-Mail: henden@fenfak.ege.edu.tr

Private address:
162 Sokak, 11/7
35040 Bornova

Domains of application:
Development of analytical methods; environmental studies and analysis; flow injection analysis

Applied methods:
Spectroscopy (atomic absorption, atomic fluorescence, molecular emission cavity analysis); voltammetry (puls, cyclic, anodic stripping); chromatography (gas chromatography and HPLC)

Membership:
IUPAC

KARTAL, Senol, Dr.
Department of Chemistry
Erciyes University
38039 Kayseri
Phone: +90 352 4374933
Fax: +90 352 4374933

Private address:
Universite Lojmanlari B-Blok. No. 58
38039 Melikgazi-Kayseri

Applied methods:
Atomic absorption spectrometry; UV-vis spectrometry; x-ray spectrometry

---

KENAR, Adnan, Dr.
Department of Chemistry
Ankara University

06100 Ankara
Phone: +90 312 2126720
Fax: +90 312 2232395
E-Mail:
kenor@science.ankara.edu.tr

Private address:
Yolac Sokak 3/12 Yenimahalle
Ankara

Domains of application:
Titrations in non-aqueous media; ion-selective electrodes; structure of metal complexes

Applied methods:
Potentiometry; x-ray spectroscopy

KILIG, Esma, Dr.
Department of Chemistry
Ankara University
06100 Ankara
Phone: +90 312 2126720
Fax: +90 312 2232395
E-Mail: ekilic@science.ankara.edu.tr

Private address:
Guzalton Sokak 10/4; 06570 Ankara

Domains of application:
Titrations in non-aqueous media; determination of equilibria constants; ion-selective electrodes

Applied methods:
Potentiometry; spectrophotometry

NEYLAN, Dirilgen, Dr.
Department of Chemistry
Bogazigi University
80815 Bebek-Istanbul
Phone: +90 212 2631816
Fax: +90 212 2742467
E-Mail: dirilgen@boun.edu.tr

Domains of application:
Environmental analytical chemistry, archeometry

Applied methods:
Atomic absorption analysis; NMR; GC; IR

OMR, Yavuz, Assoc. Prof. Dr.
Department of Chemistry
University of Dicle
21280 Diyarbakir
Phone: +90 412 2488550

Private address:
Dicli Universitesi Lojmanlari, D8-4
Diyarbakir

ONER, A. Nur, Prof.
Chemistry Department
Faculty of Art and Science
Kurupelit
55139 Samsun
Phone: +90 362 4576022
Fax: +90 362 4576081
E-Mail:
nuronar@vm.baum.omu.edu.tr

Private address:
Universite Lofmanlari 6
55139 Samsun

Domains of application:
Research work on various subjects using chromatographic, spectroscopic and electrochemical techniques

Membership:
International Association on Water Quality

OZDEMIR, Mustafa, Dr.
Department of Chemistry
Karadeniz Technical University
61080 Trabzon
Phone: +90 462 2487017
Fax: +90 462 2487344

Private address:
Universite Lojmanlari 24/7
61100 Trabzon

Turkey

Domains of application:
Analytical methodology in new synthetic chemicals; analysis of natural materials; revision of some analytical methods

Applied methods:
Spectrophotometric methods; titrimetric methods; chromatographic methods

Membership:
Turkish Society of Chemistry; IUPAC; International Association of Educator for Worl Peace

---

OZYORUK, Haluk, Dr.
Departmant of Chemistry
Hacettepe University
06532 Beytepe-Ankara
Phone: +90 312 2352596
Fax: +90 312 2352596

Private address:
Drulumak Sokak 59/8 Kocotepe
06648 Ankara

Domains of application:
Electroanalytical chemistry; amperometric enzyme electrodes; electroorganic chemistry

Applied methods:
Voltammetric; polarographic; coulometric; potentiometric; conductimetric; amperometric methods

---

PEKMEZ, Kadir, Dr.
Department of Chemistry
Hacettepe University
06532 Beytepe-Ankara
Phone: +90 312 2352596
Fax: +90 312 2352596
E-Mail: pekmez@eti.cc.hun.edu.tr

Private address:
Yargi-2 Sitesi 3 Cadde 40
06370 Ankara

Domains of application:
Electroanalytical chemistry; conducting polymers; electrocatalysis; electroorganic chemistry; industrial electrochemistry

Applied methods:
Voltammetric; coulometric; potentiometric; conductimetric; amperometric methods

---

PEKMEZ, Nuran, Dr.
Department of Chemistry
Hacettepe University
06532 Beytepe-Ankara
Phone: +90 312 2352596
Fax: +90 312 2352596
E-Mail: pekmez@eti.cc.hun.edu.tr

Private address:
Yargi-2 Sitesi 3 Cadde 40
06370 Ankara

Domains of application:
Electroanalytical chemistry; conducting polymers; electrocatalysis; electroorganic chemistry; industrial electrochemistry

Applied methods:
Electrochemical methods

SALIH, Bekir, Assoc. Prof.
Department of Chemistry
Hacettepe University
06532 Beytepe-Ankara
Phone: +90 312 2352596
Fax: +90 312 2352596
E-Mail: bekir@eti.cc.hun.edu.tr

Private address:
Hosdere Cad. Portakal Cicegi Sok.
41/3
06540 Ankara

Domains of application:
Heavy metal removal from waste

water; bioaffinity chromatography; dye affinity and metal chelate affinity chromatography; pesticide removal using polymeric sorbents; speciation of metal ions which have different oxidation state

Applied methods:
Speciation of Se in human serum using HPLC-AAS; removal of heavy metals ions with dye attached poly (HEMA-EGDMA) microbeads; degradation of phenolic species using microorganisms; polymeric base chromatographic techniques

---

SAVASCI, Sahin, Dr.
Cumhuriyet University
58140 Sivas
Phone: +90 346 2261527

Domains of application:
Education/research

SERDAR, Ates, Prof.
Department of Chemistry
Hacettepe University
06532 Beytepe-Ankara
Phone: +90 312 2351951
Fax: +90 312 2352596
E-Mail: ates@eti.cc.hun.edu.tr

Private address:
Ahmet Rasim Sokak No. 8/8
06550 Sankaya-Ankara

Domains of application:
Electrochemistry; photochemistry; spectroscopic; fiber optic sensors

Applied methods:
Fluorescence; phosphorescence; chemiluminescence; chromatographic separations; photochemical analysis

Membership:
Turkish Chamber of Chemical Engineers; The Chemical Society of Turkey

---

SOMER, Guler, Dr.
Department of Chemistry
Gazi University
Kimyu Bolum
06500 Ankara
Phone: +90 312 2122900
Fax: +90 312 2122279
E-Mail:
kimya@quark.fef.gazi.edu.tr

Private address:
Esat Cad. 42-15
06660 Ankara

Domains of application:
Electroanalytical chemistry; preparation of ion selective electrodes; trace element determination in various media; various investigations for selenium determination

Applied methods:
Voltammetric (Ads. Strip., anodic and cathodic stripping); polarographie; coulometric; spectrophotometric methods

Membership:
WPAC; IUPAC; National Geography Society; Turkish Chemical Society

Awards:
Food Foundation Schorlarship 1969-1970

SOYLAK, Mustafa, Dr.
Fen-Edebiyat Fakultesi
Erciyes University
Kimya Bolumu
38039 Kayseri
Phone: +90 352 4374901
Fax: +90 352 4374933

Private address:
Fatih Mh. Gengler Ap. Kat 3

38020 Kayseri
Domains of application:
Determinations of trace metals from various matrices after preconcentration procedures on some adsorbents

Applied methods:
Atomic absorption spectrometric; determination; spectrophotometric; determinations

---

TULLIN, Tulay, Prof. Dr.
Chemistry Department
Faculty of Science and Letters
Maslak
80626 Istanbul
Phone: +90 212 2853249
Fax: +90 212 2856386
E-Mail: tulun@sanyer.cc.itu.edu.tr

Private address:
I.T.U. Lajmanlari B.II, 2/12
80626 Istanbul

Domains of application:
Archeometry; polyelectrolytes and polyelectrolyte complexes; trace element analysis; surface coatings by sol-gel technique

Applied methods:
X-ray techniques (SEM, XRD, XRF, EDA); conductometry and potentiometry; atomic absorption spectrophotometry; flame photometry; infrared spectroscopy

Membership:
Turkish Chemical Society

TURKER, A. Rember, Prof. Dr.
Science and Art Faculty
Gazi University
Tehnih Ohallar
06500 Ankara
Phone: +90 312 2122900
Fax: +90 312 2122279

Domains of application:
Atomic absorption spectroscopy; atomic emission spectroscopy; trace element analysis; studying on preconcentration methods

Applied methods:
Analysis of geological samples, alloys, anodic slurries and biological samples

---

TUTEM, Esma, Dr.
Department of Chemistry
Faculty of Engineering
Avcilar Campus
34850 Istanbul
Phone: +90 212 5911998
Fax: +90 212 5911997

Private address:
Sirmakep Sok. Oxlem Apt B, Blok. 22
81480 Kucutyali-Istanbul

Domains of application:
Development of spectrophotometric methods of analysis for trace elements, for environmentallly important elements and reducing agents, for biologically important compounds; heavy metals and organic pollutants removal from aqueous solution

Applied methods:
Molecular and atomic spectrometry (UV-VIS, spectrophotometry, AAS); HPLC

YILDIZ, Attila, Dr.
Department of Chemistry
Hacettepe University
06532 Beytepe-Ankara
Phone: +90 312 2352596
Fax: +90 312 2352596
E-Mail: yildiz@eti.cc.hun.edu.tr

Private address:
MESA Guneg Sitesi Cayhone Sok

3/11
Ankara

Domains of application:
Electroanalytical chemistry; sensors based on conducting polymers; electrocatalysis; electroorganic chemistry

Applied methods:
Electrochemical methods

Awards:
Alexander von Humboldt Scholarship, 1973; DAG Fellowship, 1987; Fullbright Fellowship, 1986

ZIYADANOGULLARI, Bern,
Assist. Prof.
Department of Chemistry
University of Dicle
21280 Diyarbakir
Phone: +90 412 2488212

ZIYADANOGULLARI, Recep,
Prof. Dr.
Analytical Chemistry Department
University of Dicle
21280 Diyarbakir
Phone: +90 412 2488212
Fax: +90 412 2488039
E-Mail: fen@kim.at.tr

Domains of application:
Hydrometallurgy; atomic spectrometry; fuel; environmental; science separation techniques; extraction

Applied methods:
ICP-AAS; leaching; liquid-liquid extraction; sulfurization

Names

A

B

C

D

E

F

G

H

I

J

K

L

M

N

T

U

V

W

Y

Z

The manufacturer's authorised representative in the EU is Springer Nature Customer Service Centre GmbH, Europaplatz 3, 69115 Heidelberg, Germany. If you have any concerns regarding our products, please contact ProductSafety@springernature.com

Printed and bound by CPI Group (UK) Ltd, Croydon, CR0 4YY
01/10/2026
02225951-0008